# Lecture Notes in Control and Information Sciences

*For further listing of published volumes please turn over to inside of back cover.*

# Lecture Notes in Control and Information Sciences

Edited by A.V. Balakrishnan and M. Thoma

## 37

Rüdiger Schmidt

# Advances in Nonlinear Parameter Optimization

Springer-Verlag Berlin Heidelberg GmbH 1982

**Author**
Dr. Rüdiger Schmidt
Universität Dortmund
Postfach 500 500
D-4600 Dortmund 50

AMS Subject Classification (1980): 65 K 05, 65 K 10, 65 F 20, 65 F 35, 15 A 03, 15 A 06, 15 A 09, 15 A 18, 15 A 60, 49 D 37, 93 B 30, 93 C 15

ISBN 978-3-540-11396-6     ISBN 978-3-540-39080-0 (eBook)
DOI 10.1007/978-3-540-39080-0

2061/3020-543210

PREFACE

For the treatment of a problem from control theory it was necessary for the author to solve a nonlinear parameter optimization problem formulated as a nonlinear least squares problem. Well-known numerical methods for its solution failed. This was the starting point for an intensive examination of the well-known numerical methods for the solution of nonlinear least squares problems. In the course of these examinations it turned out that these methods are not free from partly severe disadvantages so that it became necessary to develop a new numerical method which does not show these disadvantages.

One main interest of these Lecture Notes is the presentation of this new method together with some of the well-known numerical methods in an uniform frame so that it will be possible to recognize the relations and differences of the methods. This is especially important for all those who are not very familiar with these methods. Besides, numerical examples demonstrate the performance of these methods so that all those who have to solve nonlinear least squares problems can consider carefully which of the presented numerical methods is most appropriate to the solution of the actual problem under consideration.

As nonlinear least squares problems arise in many very different fields it is convenient to point out the underlying principles and ideas without relating to special applications. The range of the occuring applications is immense so that it is only possible to consider some applications by example. Here these examples are taken from control theory but they can easily be substituted by others.

ACKNOWLEDGEMENTS

The author is deeply indebted to Prof. Dr. Kiendl, Lehrstuhl für Elektrische Steuerung und Regelung at University of Dortmund, West Germany, who has inspired the study of nonlinear least squares problems and who has supported the writing of these Lecture Notes. It was Prof. Dr. H. Kiendl who encouraged the author and who helped him with extensive discussions and detailed criticism.

The manuscript typing, from the first draft to the final version, was undertaken by Mrs. G. Kasimir. The author offers his sincerest thanks to her for this tedious and troublesome work which she has done with extreme care. Moreover the author thanks Mrs. D. Bandow and Mrs. G. Marx for drawing the figures, tables, and logic diagrams.

TABLE OF CONTENTS

# 1. INTRODUCTION

In many very different fields it is necessary to represent a great number of data points in an easily understandable way. Usually, such data points are dependent on one or more independent variables. If the data points are only dependent on one independent variable it is possible to plot the data points and to draw a curve through them. Then this curve is the graphical representation of the data points. If the data points do not only depend on one independent variable it is not so easy to produce a graphical representation for them. In this case it is necessary to look for other possibilities of the representation, for instance a functional form. The substitute of the data points by a functional form is only convenient if the data points are well represented by it. A functional form is also of interest if the data points are to be used for computations on a digital computer because it is not necessary to store the data points - which can be a very great number - but only the functional form as a representation for them. Moreover an easy interpolation between data points is possible with the help of a functional form.

For these reasons we must enter into the question how to obtain such a functional form. Usually, a class of functions is selected - for instance the class of polynomials, of exponential functions, of trigonometric functions, and so on. If we assume that each element of a selected class has a parametric representation then the individual functions are characterized by different values for the parameters. As the functional form shall be as good a substitute as possible of the data points we must determine that function out of the functions of the selected class which fits the data points best in the sense of an error criterion. As this best function is characterized by certain values for the parameters we must fix these parameters in an appropriate way. This can be done by optimizing the error criterion with respect to the parameters. The determination of a functional form as representation for the data points in this way is usually called curve fitting.

It is obvious that there are two degrees of freedom, namely the class of functions to be selected and the error criterion to be chosen. By selecting a special class of functions and choosing a special criterion for judging the quality of the curve fitting these two degrees of freedom can be used to make the curve fitting problem easy from a computational point of view. To see this we consider the following problem: Let $f_j$, $j = 1, \ldots, n$, be any functions of one or more independent variables. Then consider the class of functions f given by

$$f = x_1 f_1 + x_2 f_2 + \ldots + x_n f_n \, . \tag{1/1}$$

This class of functions has the property that eq. (1/1) is linear in the parameters $x_j$, $j = 1, \ldots, n$. Let the given data points be $y_i$, $i = 1, \ldots, m$, and denotate the corresponding values of the function f by $f_i$. If we choose the sum of squares

$$Q = \sum_{i=1}^{m} (y_i - f_i)^2 \tag{1/2}$$

as error criterion the values of the parameters $x_j$ from eq. (1/1) can be determined analytically by minimizing Q from eq. (1/2).

If the class of functions is given by eq. (1/1) and if the error criterion (1/2) is used we speak of a linear least squares problem. This is a linear problem because the set of equations from which the parameters are determined is linear in the parameters which is a consequence of the selected class of functions (1/1) and of the applied error criterion (1/2). For this linear least squares problem an analytical solution can be given. However the computation of this solution on a digital computer can cause severe problems. A detailed treatment of the linear least squares problem with all its special aspects can be found for instance in [1] where also numerical problems are taken into consideration.

If the applied error criterion is given by eq. (1/2) but the class of functions is not given by eq. (1/1) we have a so-called nonlinear least squares problem. It is a nonlinear problem because the set of equations from which the parameters are determined is nonlinear in the parameters. In general this problem cannot be solved analytically so that we are depending on appropriate numerical methods for its solution. For the solution of these nonlinear least squares problems iterative methods are of special interest.

Because of the two degrees of freedom in curve fitting problems we can avoid nonlinear least squares problems in principle. But it can happen that the given data points cannot be fitted as well as desired by a function from the class (1/1). Then it can be necessary to consider also classes of functions f which are not linear in the parameters. That means curve fitting problems may lead to nonlinear least squares problems.

Such nonlinear least squares problems cannot be avoided as a matter of principle if we are not free to select the class of functions as this class is given by the physical nature of the process where the data points come from. Often we are familiar to a certain extent with the laws which govern the process under consideration. Then it is possible to derive equations for instance from physical, chemical, economic laws, and so on which describe the performance of the process under consideration. By applying these laws the class of functions is selected. If all numerical values for the occuring parameters are known the function describing the process is determined. But often some or all numerical values of the parameters are not known so that the problem remains to determine these numerical values. If it is possible to measure some quantities of the process under consideration we have the chance to determine the unknown numerical values of the parameters by the aid of these measurements. For this purpose we demand that the measured values are fitted as well as possible by a function f out of the class of selected functions. This can be achieved by introducing an appropriate error criterion and optimizing it. By this procedure we obtain the numerical values of the parameters we are looking for. This processing is usually called model fitting. The function f describing the process under consideration is called a mathematical model or simply a model. In model fitting we have only one degree of freedom, namely the error criterion which is used. Often the sum of squares from eq. (1/2) is utilized. As the model is usually nonlinear in the parameters model fitting yields nonlinear least squares problems in general.

For doing model fitting we perform some measurements in order to obtain data points which have to be fitted. From this procedure we obtain the values for the parameters occuring in the model. These parameters usually have a physical significance if the model is a correct one. Therefore it makes sense to ask for the true values of these parameters. As the accuracy of the measurements performed at the process under consideration is never perfect it cannot be expected that the values of the parameters determined by the solution of the model fitting problem are identical with their true values. But it is desirable that they are close to their true values. Another aspect which is important in model fitting concerns the inevitable random errors during the registration. If we have one process and if we perform a series of measurements under the same conditions several times the measured values will be different each time due to the random errors. In general we can expect that the measured values fluctuate around their mean values. If we perform the model fitting for each series of measurements we will obtain different values for the parameters each time. It is desirable that these values also fluctuate around their mean values and that they do not vary extremely from one series of measurements to another.

From these considerations the desire stems that the parameters are close to their true values and approximately identical for several series of measurements performed at the same process under the same conditions. That means that we look for parameters with supplementary statistical properties. As far as that is concerned this is an extension of the model fitting problem. Usually, this problem is called model estimation problem. It is obvious that the model estimation and the model fitting problem are similar but that the model estimation problem will be more complicated. As the selection of the class of functions is identical in model fitting and model estimation we also have one degree of freedom in the latter case, namely the error criterion. If we use the sum of squares as error criterion the model estimation problem also yields a nonlinear least squares problem. A detailed treatment of the model estimation problem with its statistic properties is given for instance in [2].

The problems of curve fitting, model fitting, and model estimation are intimately related to the solution of least squares problems. Besides these problems there is a lot of other problems from various fields which are no least squares problems as a matter of principle but which can be solved by means of an appropriate formulation as a least squares problem. That means the originally given problem is substituted by another problem which is a least squares problem. Here we want to give some examples to illustrate this.

Often we have the problem to look for a mathematical model of a given dynamic system, for instance in the form of transfer function or sets of differential equations. Such mathematical models are of interest for analytic investigations or for simulations on computers. If the dynamic system is not very complex and if we have a good idea of the processes in the interior of the system it is possible to derive such a desired mathematical model of the dynamic system from the laws which govern the processes in the interior. If we know all numerical values of the parameters occuring in the mathematical model we have reached our goal. If not all numerical values of the parameters are known this problem leads to a model fitting problem if measurements may be performed. Often we do not succeed in deriving a mathematical model of the dynamic system from the laws that govern the processes. In this case we can try to determine a mathematical model only from measurements of the inputs and outputs of the dynamic system. The proceeding which is called identification may be as follows: We select a class of functions which shall describe the mathematical model of the dynamic system where some parameters are still undetermined. Then we demand that the measured outputs of the dynamic system are fitted as well as possible in the sense of an error criterion by the outputs obtained from the mathematical model when applying the measured inputs. By optimizing the error criterion we can fix the values for the parameters occuring in the functions of the selected class. If we utilize the sum of squares as error criterion this problem leads to a nonlinear least squares problem.

Another problem which can also be solved by means of a formulation as a least squares problem is the so-called simplification of mathematical models. Considering this kind of problems we assume that we already know a mathematical model, for instance a transfer function. We want to replace this model by another one which is simpler somehow but which is similar to the originally given model. There is a lot of methods for the solution of this problem. A survey is given for instance in [3] and [4]. To achieve the required similarity of the simplified model to the originally given model we can ask for such a model which fits for instance the step response, impulse response, or frequency response of the given model as well as possible in the sense of an appropriate error criterion. If we select a certain class of functions for the description of the simplified mathematical model where some parameters are not yet fixed the problem of model simplification formulated in this way also yields a nonlinear least squares problem.

One main interest in control theory is the design of controllers for a given plant so that the closed-loop control system shows a desired performance. One possible design procedure is the so-called Bode design whose details are explained for instance in [5]. The philosophy of this design procedure is to transform some specifications usually given in the time-domain for the closed-loop control system to specifications for the Bode magnitude and phase angle plots of the open-loop control system. In general the given plant will not satisfy these specifications so that it is necessary to introduce a compensator. If a class of functions for the description of this compensator is selected - for instance those describing lag-lead compensators - where some parameters are still undetermined it is possible to fix these parameters in such a way that the required Bode magnitude and phase angle plots are fitted as well as possible in the sense of an error criterion. We recognize that the design problem formulated in this way also yields a nonlinear least squares problem. This processing is also possible if we use other design procedures instead of the Bode design procedure.

As last example of this type we consider a problem which arises in optimal control theory. There the following problem is of interest: Given a mathematical model for a dynamic system. Determine such a control function as input of the mathematical model so that a dynamic variable shows a desired performance. If we select a class of functions as control functions where some parameters are still undetermined we can demand that the control function is determined so that the desired performance of the dynamic variable is fitted as well as possible in the sense of an error criterion by the dynamic variable when utilizing a certain control function out of the selected class as input. Formulated in this way a control function can be determined by the solution of a nonlinear least squares problem.

These presented examples show that there really is a lot of problems which are no least squares problems as a matter of principle but the solution of which can be achieved by an appropriate formulation as a nonlinear least squares problem. Often such a formulation is the only practicable way to solve such problems because the primarily given problems cannot be treated analytically or because they can only be solved numerically with great effort in comparison with a solution resulting from a nonlinear least squares problem.

These few examples show that it is necessary to study the solution of nonlinear least squares problems thoroughly. In general, their solution cannot be determined analytically but only by the aid of a numerical method implemented on a digital computer. Therefore it is very important to have the disposal of stable, reliable, and efficient numerical methods. There is a lot of well-known methods which are appropriate to the solution of nonlinear least squares problems. But all these methods have some disadvantages which are more or less troublesome. In order to recognize these disadvantages we have to review the most frequently used methods which is done in Chapter 3 of these Lecture Notes. Thereupon in Chapter 4 a new iterative method for the solution of nonlinear least squares problems is proposed and discussed in details. This method is free of the disadvantages of the most frequently used methods. Finally in Chapter 7 several examples for nonlinear least squares problems are presented. Their solutions are determined by the aid of several versions of the proposed new method and of other methods in order to compare these methods among one another.

In Chapter 2 the nonlinear least squares problem to be treated here is formulated mathematically as a nonlinear minimization problem. Well-known iterative numerical methods for the solution of this nonlinear minimization problem are presented in Chapter 3. Thereby we distinguish between numerical methods for the minimization of arbitrary functions depending on several variables and numerical methods which take care of the special structure of the function to be minimized in the case of a nonlinear least squares fit. A critical judgement of the presented methods leads to some requirements which have to be satisfied by an always applicable numerical method for the solution of nonlinear least squares problems. At the top of Chapter 4 these requirements are formulated. Thereupon a new iterative numerical method is derived which complies with the posed requirements. An essential element in this proposed method is an intelligent selection of some vectors of the Jacobian matrix in the course of the iteration. As to this selection of the vectors there are some degrees of freedom. They are removed by the introduction of convenient criteria for the selection of these vectors. Thereby the requirement that the numerical effort is as small as possible plays an important part. As it is necessary to invert a matrix in the course of the iteration the question concerning the condition of this matrix is discussed in details. Furthermore criteria for the finishing of the iteration are derived and a numerical method for the solution of the so-called one-dimensional search is described.

The linear least squares problem can be considered as a special case of the nonlinear least squares problem. In Chapter 5 it is shown how the proposed new method can be applied to the solution of this problem and what kind of simplifications results in this case. Furthermore a statement can be made about the way of convergence of the proposed method. Moreover the criteria for the selection of the vectors introduced in Chapter 4 are investigated in this special case of a linear least squares problem.

To start any iterative method it is necessary to have a starting point. The choice of such a starting point is very important for the performance of an iterative method. Therefore the question how to obtain a "good" starting point is treated in Chapter 6. It is shown that an appropriate mathematical representation of a function can reduce this problem considerably. In particular this is true if we have to solve a nonlinear least squares problem arising from a real technical system. For such systems we often have informations or ideas of the order of magnitude of some or all parameters.

Examples for the application of the presented iterative methods for the solution of nonlinear least squares problems are given in Chapter 7. The examples considered there are taken from control theory. They deal with the following problems: identification of linear time-invariant dynamic systems, simplification of mathematical models of linear and nonlinear dynamic systems, and determination of control functions. As a matter of principle these problems are no nonlinear least squares problems but by an appropriate formulation they can be transformed to such problems and can then be solved by the methods presented here. For this it is necessary to derive appropriate functions and relations for their derivatives with respect to the parameters. After having prepared the problems in such a way it is possible to study the performance of several versions of the proposed new method for various starting points. Moreover a comparison of the new method with other well-known methods is performed. These investigations demonstrate the power of the proposed new method.

References:

[1] Lawson, Ch.L.; Hanson, R.:
    Solving Least Squares Problems
    Englewood Cliffs, N.J.: Prentice-Hall, Inc. (1974).
[2] Bard,Y.:
    Nonlinear Parameter Estimation
    New York and London: Academic Press (1974).
[3] Bosley, M.J.; Lees, F.P.:
    A Survey of Simple Transfer-Function Derivations from High-Order State Variable
    Models
    Automatica 8, 765-775 (1972).
[4] Gwinner, K.:
    Vereinfachung von Modellen dynamischer Systeme
    Regelungstechnik 10, 325-333 (1976).
[5] DiStefano, J.; Stubberud, A.R.; Williams, I.J.:
    Theory and Problems of Feedback and Control Systems
    New York: McGraw-Hill, Inc. (1967).

## 2. FORMULATION OF THE PROBLEM

Given a finite dimensional set of fixed values $y_i$, $i = 1, \ldots, m$. With these given values $y_i$ we build up the vector

$$\underline{y} = \begin{pmatrix} y_1 \\ y_2 \\ \cdot \\ \cdot \\ \cdot \\ y_m \end{pmatrix} . \qquad (2/1)$$

In the sequel we assume that the components of the vector $\underline{y}$ are all real-valued. Therefore the vector $\underline{y}$ is an element of the m-dimensional space where the components of the vector $\underline{y}$ are elements in the field of all real numbers. This space shall be denotated by $\mathbb{R}^m$.

The given fixed values $y_i$ are to be fitted by functions $f_i$, $i = 1, \ldots, m$ as well as possible in the sense of an appropriate error criterion. We assume that the structure of the functions $f_i$ is given but that the functions $f_i$ still depend on some parameters $x_j$, $j = 1, \ldots, n$ which are to be determined. For simplification of the notation we build up the vector

$$\underline{x} = \begin{pmatrix} x_1 \\ x_2 \\ \cdot \\ \cdot \\ \cdot \\ x_n \end{pmatrix} \qquad (2/2)$$

which we call the parameter vector.

Furthermore we demand that the components $x_j$ of the vector $\underline{x}$ are all real-valued. Then the vector $\underline{x}$ is an element of the space $\mathbb{R}^n$.

It is convenient to build up a vector $\underline{f}$ which contains the functions $f_i$, $i = 1, \ldots, m$. This vector $\underline{f}$ depends on the parameter vector $\underline{x}$, i. e.

$$\underline{f} = \begin{pmatrix} f_1(\underline{x}) \\ f_2(\underline{x}) \\ \cdot \\ \cdot \\ \cdot \\ f_m(\underline{x}) \end{pmatrix} = \underline{f}(\underline{x}) . \qquad (2/3)$$

The function $\underline{f}$ is a mapping from the n-dimensional space $\mathbb{R}^n$ to the m-dimensional space $\mathbb{R}^m$. This property can be symbolized by

$$\mathbb{R}^n \xrightarrow{\ f\ } \mathbb{R}^m. \tag{2/4}$$

For the function $\underline{f}$ we admit linear mappings as well as nonlinear mappings. We assume merely that the mapping $\underline{f}$ is twice continuously differentiable with respect to $\underline{x}$. By the mapping $\underline{f}$ we obtain a value $f_i$ corresponding to each given value $y_i$. The resulting value $f_i$ still depends on the parameter vector $\underline{x}$. According to the choice of the components of the parameter vector $\underline{x}$ which are actually present the fixed values $y_i$ are "better" or "worse" fitted by the values $f_i(\underline{x})$.

The above-mentioned verbal demand concerning as good a fit as possible of the given fixed values $y_i$ must be transformed to a mathematical relation which can be evaluated. For that purpose we introduce the error vector

$$\underline{\bar{e}} = \underline{y} - \underline{f} \ . \tag{2/5}$$

On the basis of this error vector we are able to judge the goodness of the fit if we introduce an appropriate error criterion. In principle, there is a lot of possibilities of choosing such an error criterion. An often used error criterion is the so-called sum of squares, which is given by

$$Q' = \underline{\bar{e}}^T\underline{\bar{e}} = \|\underline{\bar{e}}\|^2 \ . \tag{2/6}$$

The vector $\underline{\bar{e}}^T$ is a row vector which is the transpose of the column vector $\underline{\bar{e}}$. By $\|\ \|$ we denote the Euclidean norm in the space $\mathbb{R}^m$.

Besides the error criterion $Q'$ which is used in a slightly modified way in the sequel there are some others [1], for instance the criterion

$$Q_1 = \sum_{i=1}^{m} |y_i - f_i(\underline{x})| \tag{2/7}$$

and the criterion

$$Q_2 = \max_{1 \le i \le m} |y_i - f_i(\underline{x})| \ . \tag{2/8}$$

The error criterion $Q_2$ leads to the so-called discrete Chebyshev approximation [2]. The numerical treatment of the criteria (2/7) and (2/8) is very difficult - especially in the case of nonlinear functions $f_i(\underline{x})$ - and is not the subject of these Lecture Notes.

Returning to the error criterion $Q'$ we can write for it

$$Q' = (\underline{y}-\underline{f})^T(\underline{y}-\underline{f}) = (\underline{y}-\underline{f}(\underline{x}))^T(\underline{y}-\underline{f}(\underline{x})) = Q'(\underline{x}) \tag{2/9}$$

if we use the relation (2/5).

The demand on as good a fit as possible of the given fixed values $y_i$ can now be satisfied by determining that value $\underline{x}^*$ of the parameter vector $\underline{x}$ for which the function $Q'(\underline{x})$ becomes minimal, i. e.

$$\underline{x}^* = \min_{\underline{x}} Q'(\underline{x}) . \qquad (2/10)$$

If we determine $\underline{x}^*$ by evaluating eq. (2/10) we call the resulting parameter vector $\underline{x}^*$ a fit in the sense of the least sum of squares. Shortly we call it a least squares fit. If we are able to solve eq. (2/10) we find a parameter vector $\underline{x}^*$ which yields as good a fit as possible of the given fixed values $y_i$ in the sense of the error criterion (2/6).

Before looking for possibilities of solving eq. (2/10) we want to generalize the error criterion (2/6) a little. Often it happens that some components of the error vector (2/5) are to obtain a greater influence on the sum of squares compared with the others. We can accomodate to these circumstances by modifying the error criterion (2/6). Instead of the error criterion $Q'$ from eq. (2/6) we consider the modified error criterion

$$Q = \underline{\bar{e}}^T \underline{P} \, \underline{\bar{e}} . \qquad (2/11)$$

We demand that $\underline{P}$ is a real positive definite symmetric matrix. Eq. (2/11) represents a very general form of the error criterion. The above-mentioned case of the different influence of the individual errors on the sum of squares is enclosed in this general form. For that purpose we have to choose a diagonal matrix $\underline{P}'$ for $\underline{P}$ in eq. (2/11) where $\underline{P}'$ consists of only real positive elements on the main diagonal

$$\underline{P}' = \begin{pmatrix} p_1' & & & \\ & \cdot & & \\ & & \cdot & \\ & & & \cdot \\ & & & & p_m' \end{pmatrix} \qquad p_i' > 0, \; i = 1, \ldots, m. \quad (2/12)$$

Such a matrix $\underline{P}'$ satisfies the general properties demanded for the matrix $\underline{P}$.

The problem to be solved can now be summarized as follows:

Given a vector $\underline{y} \in \mathbb{R}^m$ and an in general nonlinear mapping $\underline{f}(\underline{x})$. Determine that vector $\underline{x}^* \in \mathbb{R}^n$ for which

$$\underline{x}^* = \min_{\underline{x}} [(\underline{y} - f(\underline{x}))^T \underline{P}(\underline{y} - \underline{f}(\underline{x}))] \qquad (2/13)$$

holds. As $\underline{f}(\underline{x})$ is a nonlinear function in general the problem to determine $\underline{x}^*$ is a nonlinear minimization problem.

This nonlinear minimization problem is called a nonlinear least squares problem because the function Q which is to be minimized is a sum of squares.

Often eq. (2/13) is written in the way

$$\underline{x}^* = \min_{\underline{x}} \| \underline{y} - \underline{f}(\underline{x}) \|_{\underline{p}} \tag{2/14}$$

where $\| \underline{z} \|_{\underline{p}}$ means

$$\| \underline{z} \|_{\underline{p}} = \underline{z}^T \underline{p} \, \underline{z} \; . \tag{2/15}$$

Here we do not utilize this notation.

When considering the problem of determining the vector $\underline{x}^*$ for which eq. (2/13) respectively (2/14) is valid the only interesting case is that for which the inequality

$$m \geq n \tag{2/16}$$

holds. Eq. (2/16) means that the number of given fixed values $y_i$ has to be greater than or equal to the number of the unknown parameters. Let m < n, so we have more unknown parameters than given values $y_i$ to be fitted. In this case it is always possible to determine such a parameter vector $\underline{x}^*$ for which the sum of squares vanishes. Therefore we assume for the further considerations that the inequality (2/16) is valid.

When considering nonlinear least squares problems which arise from practical applications the inequality

$$m \gg n \tag{2/17}$$

often holds (compare chapter 7).

In the sequel we shall show up several possibilities of solving the above formulated nonlinear least squares problem. Hereby we are especially interested in iterative numerical methods for its solution. A survey of such well-known iterative methods is given in the next chapter.

References:

[1] Hayes, J.G.:
    Algorithms for Curve and Surface Fitting
    Proc. of the Loughborough University of Technology Conference of the Institute of
    Mathematics and its Applications held in April 1973. Edited by D.J. Evans.
    New York and London: Academic Press (1974).
[2] Stoer, J.:
    Einführung in die Numerische Mathematik I
    Heidelberger Taschenbücher
    Berlin, Heidelberg, New York: Springer-Verlag (1972).

# 3. WELL-KNOWN METHODS FOR THE SOLUTION OF NONLINEAR LEAST SQUARES PROBLEMS

There are several possibilities of solving nonlinear least squares problems which have been formulated in chapter 2. To begin with we can look for an analytic solution of the problem (2/13). But this proceeding is only successful if the problem is very simple. Therefore in all other cases we are depending on numerical methods which are in general implemented on digital computers. Hereby we can distinguish two different kinds of methods in principle. The methods of the first kind are developed to determine the minimum or maximum of an arbitrary function depending on several variables. These methods are called universal minimizing methods. The methods of the second kind are developed to determine the minimum or maximum of a function with the special structure (2/11) which depends on several variables. These methods are called special minimizing methods.

In section 3.1 a representation of the basic ideas of a great number of universal minimizing methods is given. After that we present the most used special minimizing methods in section 3.2. Finally in section 3.3 we compare the two kinds of methods concerning their efficiency in solving nonlinear least squares problems.

## 3.1 Solution of nonlinear least squares problems with the help of universal minimizing methods

There is a lot of universal minimizing methods [1, 2, 3] which are all more or less "efficient". How "efficient" a method is can be judged by several criteria which may be very different depending on the actual point of view. Therefore a comparison which is valid for all problems under consideration cannot be performed. Possible criteria which are often used are for instance the needed computation time, the number of function evaluations, the number of iteration steps - each time for the solution of a certain test problem - , the needed core, and so on.

Depending on the actually chosen criterion, one method can be more efficient than all others. In general we can state that the so-called gradient methods are more efficient than for instance the random search methods or nonsequential factorial search methods [3]. This is especially true if we are able to determine the gradient of the function to be minimized analytically. But sometimes it happens that we must choose such a less efficient method because certain types of problems are amenable only to these methods. Moreover it may be necessary to start a minimization process with such a method to find "good" starting points for another iterative method with which we continue the minimization process. Furthermore these methods can be used if the function to be minimized exhibits radical vacillations.

Here we only consider gradient methods in more details. We call a method a gradient method if we need the knowledge of the gradient besides the knowledge of the function value in order to find the parameter vector $\underline{x}^*$ we are looking for. Higher derivatives of the function to be minimized are not to be used because their determination is quite expensive in general; especially if we have no analytical relations for them.

At first we present the basic principles which all gradient methods have in common. For that purpose we consider the following problem:

Given a real-valued twice continuously differentiable function h depending on n real variables $x_1$, ..., $x_n$. Gathering up these variables in the vector $\underline{x}$ we can write

$$h = h(\underline{x}), \ \underline{x} \in \mathbb{R}^n. \tag{3.1/1}$$

We look for that value $\underline{x}^*$ for which $h(\underline{x})$ becomes minimal. If we are looking for that value $\underline{x}^*$ for which $h(\underline{x})$ becomes maximal we consider the function $\tilde{h}(\underline{x}) = -h(\underline{x})$ and look for the minimum of the function $\tilde{h}(\underline{x})$. Therefore the case that we are looking for a minimum does not restrict the generality. For this reason we only consider the problem of determining a minimum in the sequel.

If $h(\underline{x})$ possesses only one minimum, the vector $\underline{x}^*$ is uniquely determined. If $h(\underline{x})$ possesses several minima, we are usually interested in the global minimum. All known numerical methods do not guarantee that the global minimum we are looking for is really reached. Rather it is necessary to gain a certain degree of security by supplementary measures that the global minimum is really reached.

It is known [2, 3] that the necessary condition for an extremum of $h(\underline{x})$ at $\underline{x}^*$ is the vanishing of the gradient of $h(\underline{x})$ at $\underline{x}^*$, that means

$$\nabla h(\underline{x})\Big|_{\underline{x}^*} = \begin{pmatrix} \frac{\partial h}{\partial x_1} \\ \cdot \\ \cdot \\ \cdot \\ \frac{\partial h}{\partial x_n} \end{pmatrix}\Bigg|_{\underline{x}^*} = \underline{0} \tag{3.1/2}$$

must hold. By inspecting the Hessian matrix $\underline{H}$ we can decide which kind of extremum has been found. The Hessian matrix $\underline{H}$ is given by

$$\underline{H}(\underline{x}) = \begin{pmatrix} \frac{\partial^2 h}{\partial x_1^2} & \cdot & \cdot & \cdot & \frac{\partial^2 h}{\partial x_1 \partial x_n} \\ \cdot & & & & \cdot \\ \cdot & & & & \cdot \\ \cdot & & & & \cdot \\ \frac{\partial^2 h}{\partial x_n \partial x_1} & \cdot & \cdot & \cdot & \frac{\partial^2 h}{\partial x_n^2} \end{pmatrix}. \tag{3.1/3}$$

If $\underline{H}(\underline{x}*)$ is positive definite, we have a minimum at $\underline{x} = \underline{x}*$ [2, 3]. Usually the matrix $\underline{H}(\underline{x})$ is symmetric because $\partial^2 h/\partial x_i \partial x_j = \partial^2 h/\partial x_j \partial x_i$ is valid in general [4].

Almost all methods for determining the minimum of the function $h(\underline{x})$ try to satisfy the necessary condition (3.1/2). In general eq. (3.1/2) is a nonlinear equation of the variables $x_1, \ldots, x_n$. The analytical solution of this nonlinear equation is only possible in a few very simple cases. Almost all practical problems cannot be solved analytically. Therefore it is necessary to look for a numerical method which yields a value $\underline{x}*$ that satisfies eq. (3.1/2). In order to find such a value $\underline{x}*$ it is convenient to use iterative numerical methods. These iterative methods work in a manner that they modify the vector $\underline{x}$ - beginning with a given starting vector $\underline{x}^0$ - in such a way that the value of the function $h(\underline{x})$ always decreases. The oldest method which works in this way is the so-called method of steepest descent. This method makes use of the fact that the negative gradient of $h(\underline{x})$ at $\underline{x}^k$ shows in the direction of the steepest descent of the function $h(\underline{x})$. ($\underline{x}^k$ denotes the actual value of the vector $\underline{x}$ in the stage k of the iteration.) Therefore it is evident that we certainly achieve a decrease of the function $h(\underline{x})$, if we choose as direction vector

$$\underline{r}^k = - \nabla h(\underline{x}^k) \tag{3.1/4}$$

for modifying the vector $\underline{x}^k$. With $\nabla h(\underline{x}^k)$ we denote the gradient of $h(\underline{x})$ at $\underline{x} = \underline{x}^k$.

As the gradient of $h(\underline{x})$ differs from point to point we are only sure that the function $h(\underline{x})$ decreases if we make an infinitesimal step in the direction $\underline{r}^k$. This means that it is necessary to determine the gradient very often, a processing which can be very inefficient. But there is the hope that the gradient does not vary drastically from point to point so that it will be possible to make a finite step in the direction $\underline{r}^k$. The largeness of this step is given by the so-called step-length factor. The performance of the method depends strongly on the choice of this step-length factor. If we determine it by evaluating

$$\alpha_{min} = \min h(\underline{x}^k + \alpha \underline{r}^k) \tag{3.1/5}$$

this is that step-length factor which is optimal in the sense of as great a decrease as possible of the function $h(\underline{x})$ after having fixed the direction vector $\underline{r}^k$. Therefore this method is often called optimal-gradient method. Eq. (3.1/5) means that it is necessary to determine the minimum of $h(\underline{x})$ along the straight line $\underline{x}^k + \alpha \underline{r}^k$. Along this straight line h is a function of only one variable instead of n variables. Therefore this process is called an one-dimensional search in the n-dimensional space. There are many known methods for determining the step-length factor $\alpha_{min}$ given by eq. (3.1/5) [1, 2, 3, 5]. The necessary numerical effort is not very great (compare section 4.7). After having determined $\alpha_{min}$ we obtain the next point $\underline{x}^{k+1}$ of the iteration process from

$$\underline{x}^{k+1} = \underline{x}^k + \alpha_{min} \underline{r}^k \ . \tag{3.1/6}$$

At the point $\underline{x}^{k+1}$ we determine the gradient again and the iteration loop begins once more. We continue the iteration as long as the absolute value of the gradient is greater than a given positive upper bound $\varepsilon$. If the absolute value of the gradient is less than $\varepsilon$ we stop the iteration and consider the last point $\underline{x}$ as the solution vector $\underline{x}^*$. (There are still further conditions for finishing the iteration but these are not essential in this context.) As only the necessary condition (3.1/2) is satisfied we must ascertain that we have really reached a minimum of $h(\underline{x})$ because it is still possible that we have found a saddle-point of $h(\underline{x})$. This examination can be done with the help of the Hessian matrix $\underline{H}$ from eq. (3.1/3).

Figure 3/1 shows the processing of the method of steepest descent in the case that $h(\underline{x})$ is a function of two variables $x_1$ and $x_2$. (In fig. 3/1 the step-length factor is not determined by eq. (3.1/5).)

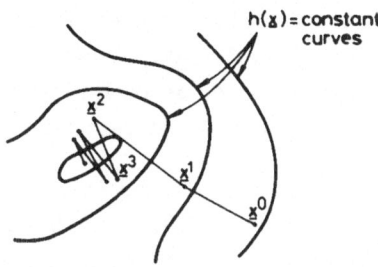

Fig. 3/1: Processing of the method of steepest descent for a function $h(x_1, x_2)$.

The advantage of the method of steepest descent is the guaranteed stability of the method that means in each stage of the iteration the function $h(\underline{x})$ decreases as long as the gradient of $h(\underline{x})$ does not vanish when utilizing an appropriate step-length factor. The disadvantage of this method is the property that it can converge very slowly to the value $\underline{x}^*$ we are looking for (see for instance fig. 3/1). Even if $h(\underline{x})$ is a quadratic function the method of steepest descent does not exactly - i. e. except for rounding errors - yields the value $\underline{x}^*$ for which $h(\underline{x})$ is minimal in a finite number of iteration steps except for special starting vectors $\underline{x}^0$ [5]. Other methods which do not solely use the negative gradient as direction vector $\underline{r}^k$ do not show this disadvantage. In many of these methods the direction vector $\underline{r}^k$ is given by

$$\underline{r}^k = - \underline{R}^k \underline{\nabla} h(\underline{x}^k) \tag{3.1/7}$$

where $\underline{R}^k$ is a real positive definite matrix. The choice of $\underline{R}^k$ as a real positive definite matrix guarantees that the resulting methods cause the function $h(\underline{x})$ to decrease in each stage of the iteration if the step-length factor is only small enough.

This can be shown if we evaluate the slope of the function $h(\underline{x})$ at $\underline{x}^k$ in the direction of the vector $\underline{r}^k$. It is given by $-(\nabla h(\underline{x}^k))^T \underline{R}^k \nabla h(\underline{x}^k)$. Because of the assumed positive definiteness of the matrix $\underline{R}^k$ the slope is negative so that the function $h(\underline{x})$ decreases in the direction of the vector $\underline{r}^k$. Otherwise the processing of these methods is equal to the one of the method of steepest descent. We call methods for which the direction vector $\underline{r}^k$ is given by eq. (3.1/7) universal gradient methods. The principal components of these methods are shown in the logic diagram 1.

By the choice of appropriate matrices $\underline{R}^k$ in eq. (3.1/7) we obtain a lot of well-known minimization methods [6]. The method of steepest descent is obtained if we choose the identity matrix $\underline{I}$ as matrix $\underline{R}^k$. From eq. (3.1/7) the often used and very successful method of Davidon, Fletcher, and Powell [7] can be obtained by choice of an appropriate matrix $\underline{R}^k$. Their method has the property that it is able to determine the minimum of a quadratic function $h(\underline{x})$ in exactly n steps except for rounding errors. A method with this property is called quadratically convergent in n steps [6]. Such methods are in general superior to the method of steepest descent concerning the number of iteration steps for determining the value $\underline{x}^*$. However the effort in programming them is much greater as well as the number of arithmetic operations in each stage of the iteration compared with the method of steepest descent.

The above described universal gradient method is qualified to solve nonlinear least squares problems. The function $h(\underline{x})$ to be minimized is the sum of squares $Q(\underline{x})$ from eq. (2/11). In order to apply the universal gradient method for the minimization of the function $Q(\underline{x})$ it is necessary to evaluate the gradient of $Q(\underline{x})$ at $\underline{x} = \underline{x}^k$. It is obtained from eq. (2/11)

$$\nabla Q(\underline{x}^k) = \nabla(\underline{\bar{e}}^T \underline{P} \,\underline{\bar{e}})\Big|_{\underline{x}^k} = (\nabla \,\underline{\bar{e}})^T\Big|_{\underline{x}^k} \underline{P} \,\underline{\bar{e}}(\underline{x}^k) +$$

$$+ (\nabla \,\underline{\bar{e}})^T\Big|_{\underline{x}^k} \underline{P}^T \underline{\bar{e}}(\underline{x}^k) \;. \tag{3.1/8}$$

Because of the assumed symmetry of the matrix $\underline{P}$, eq. (3.1/8) can be written as

$$\nabla Q(\underline{x}^k) = 2(\nabla \,\underline{\bar{e}})^T\Big|_{\underline{x}^k} \underline{P} \,\underline{\bar{e}}(\underline{x}^k) \;. \tag{3.1/9}$$

Observing eq. (2/5) the not yet defined expression $\nabla \,\underline{\bar{e}}$ is given by

$$(\nabla \,\underline{\bar{e}})\Big|_{\underline{x}^k} = -(\nabla \,\underline{f})\Big|_{\underline{x}^k} = - \left. \begin{pmatrix} (\nabla \underline{f}_1)^T \\ \vdots \\ (\nabla \underline{f}_m)^T \end{pmatrix} \right|_{\underline{x}^k} \;. \tag{3.1/10}$$

Introducing the vectors $\underline{\bar{d}}_j(\underline{x}^k)$, j = 1, ..., n according to

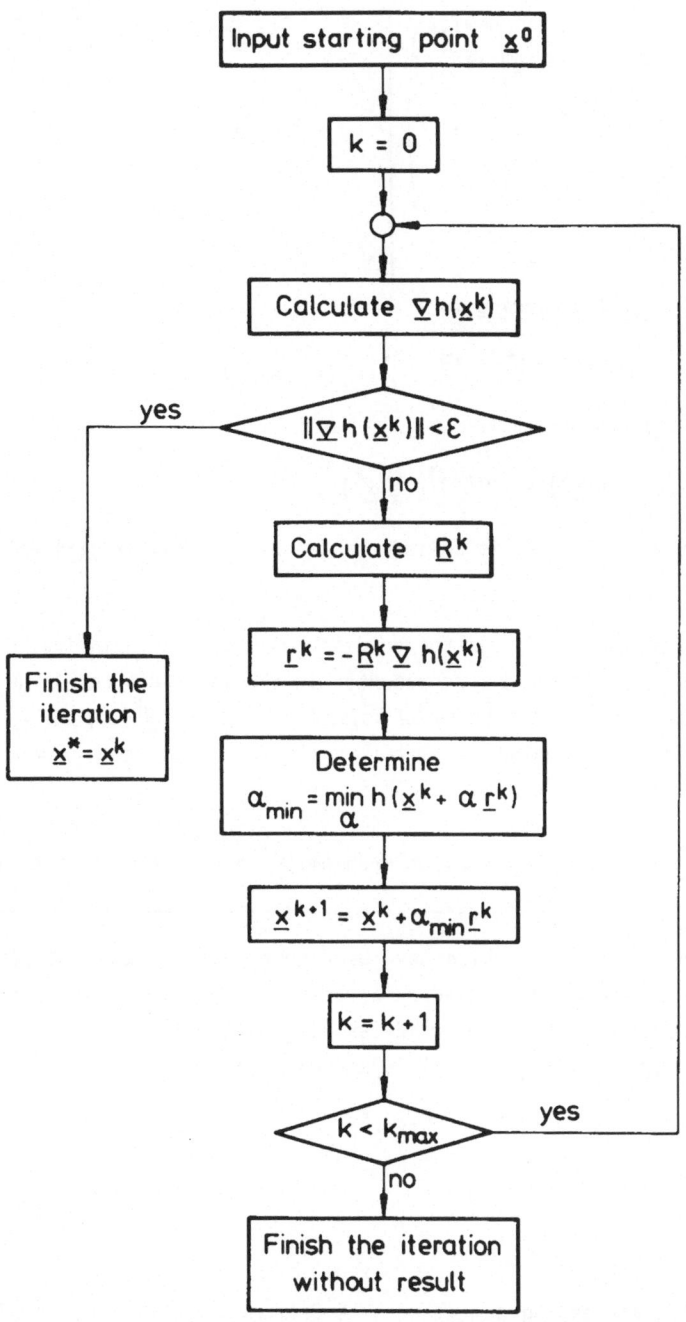

Logic diagram 1: Principal components of universal gradient methods for the minimization of a function $h(\underline{x})$.

$$\underline{d}_j(\underline{x}^k) = \begin{pmatrix} \dfrac{\partial f_1}{\partial x_j} \\ \vdots \\ \dfrac{\partial f_m}{\partial x_j} \end{pmatrix} \Bigg|_{\underline{x}^k} \quad , \; j = 1, \ldots, n \tag{3.1/11}$$

and the mxn-matrix $\underline{D}(\underline{x}^k)$ according to

$$\underline{D}(\underline{x}^k) = \underline{D}^k = (\underline{d}_1(\underline{x}^k), \ldots, \underline{d}_n(\underline{x}^k)) \tag{3.1/12}$$

the gradient of $Q(\underline{x})$ can be written as

$$\underline{\nabla}Q(\underline{x}^k) = -2(\underline{D}^k)^T \underline{P} \, \overline{\underline{e}}(\underline{x}^k) \; . \tag{3.1/13}$$

The matrix $\underline{D}^k$ from eq. (3.1/12) is called the Jacobian matrix of $\underline{f}(\underline{x})$ with respect to $\underline{x}$.

Utilizing the relation (3.1/13) for the gradient of $Q(\underline{x})$, the principal components of the universal gradient method are represented in the logic diagram 2 in case that a nonlinear least squares problem is to be solved. The matrix $\underline{R}^k$ appearing in this logic diagram has to be determined according to the just selected universal gradient method.

## 3.2 Solution of nonlinear least squares problems with the help of special minimizing methods

The special minimizing methods which have been developed for the solution of nonlinear least squares problems are based on an approximation of the function $\underline{f}(\underline{x})$. The function $\underline{f}(\underline{x})$ is expanded in a Taylor series at $\underline{x}^k$ which is truncated after the linear term, that means the function $\underline{f}(\underline{x})$ is approximated by

$$\underline{f}(\underline{x}) \approx \underline{f}(\underline{x}^k) + \underline{D}^k(\underline{x}-\underline{x}^k) \; . \tag{3.2/1}$$

To abbreviate the notation we write

$$\underline{r}^k = \underline{x} - \underline{x}^k \; . \tag{3.2/2}$$

Besides the really interesting sum of squares $Q$ from eq. (2/11) we consider another sum of squares $\overset{\approx}{Q}$ which is obtained if we utilize the approximation (3.2/1) instead of $\underline{f}(\underline{x})$ itself. This sum of squares is given by

$$\overset{\approx}{Q} = (\underline{y}-\underline{f}(\underline{x}^k) - \underline{D}^k\underline{r}^k)^T\underline{P}(\underline{y}-\underline{f}(\underline{x}^k) - \underline{D}^k\underline{r}^k)$$

$$= (\overline{\underline{e}}(\underline{x}^k) - \underline{D}^k\underline{r}^k)^T\underline{P}(\overline{\underline{e}}(\underline{x}^k) - \underline{D}^k\underline{r}^k) \; . \tag{3.2/3}$$

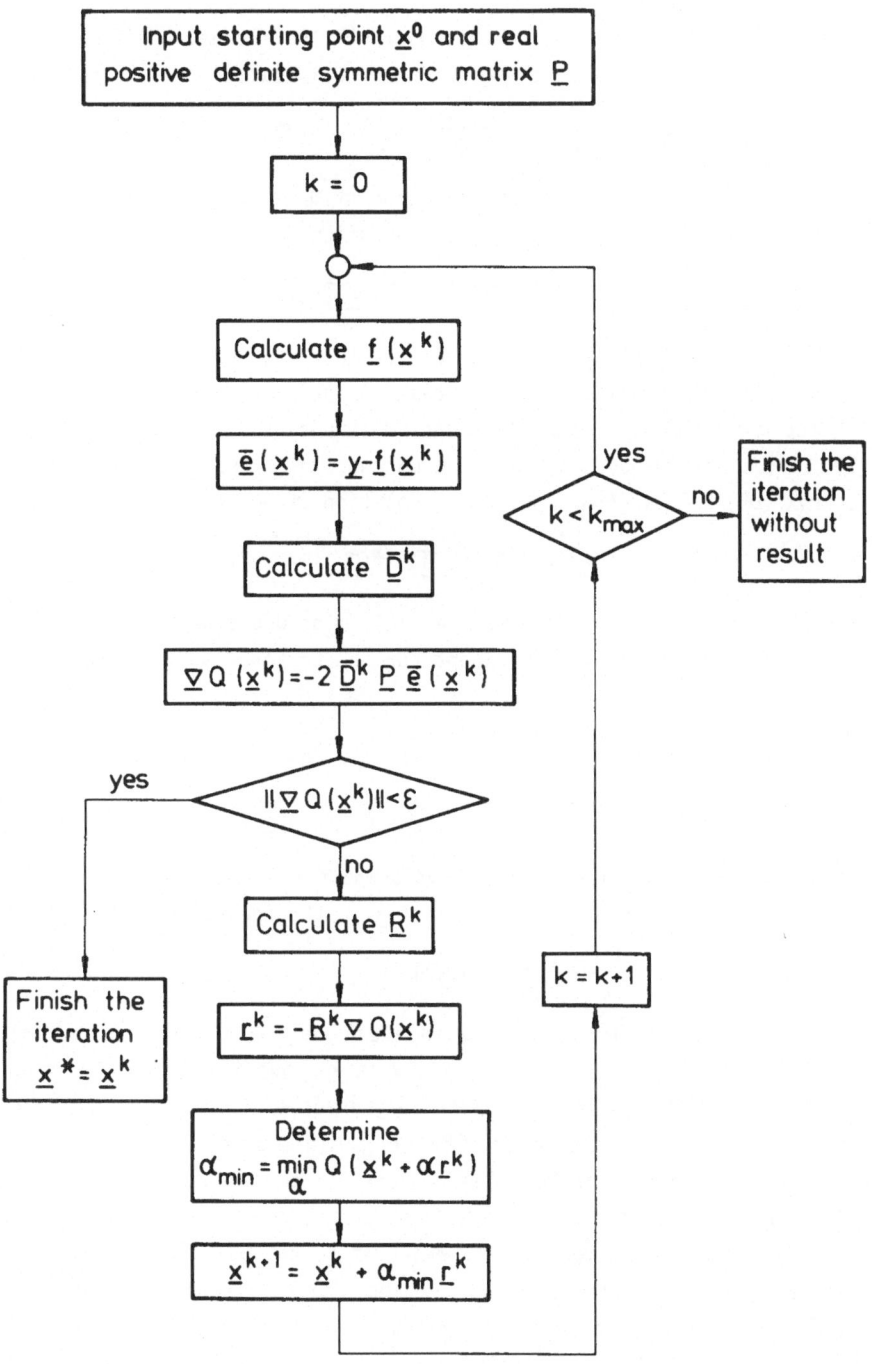

Logic diagram 2: Principal components of universal gradient methods for the solution of nonlinear least squares problems.

$\overset{\approx}{Q}$ is a function which is quadratic with $\underline{r}^k$. The extremum of $\overset{\approx}{Q}$ can be evaluated analytically. For that purpose the gradient of $\overset{\approx}{Q}$ is calculated. It is given by

$$\underline{v}\overset{\approx}{Q} = -2(\underline{D}^k)^T\underline{P}\ \overline{\underline{e}}(\underline{x}^k) + 2(\underline{D}^k)^T\underline{P}\ \underline{D}^k\underline{r}^k \quad . \tag{3.2/4}$$

The necessary condition for an extremum of $\overset{\approx}{Q}$ is the vanishing of its gradient, that means the vector $\underline{r}^k$ must satisfy the equation

$$(\underline{D}^k)^T\underline{P}\ \underline{D}^k\underline{r}^k = (\underline{D}^k)^T\underline{P}\ \overline{\underline{e}}(\underline{x}^k) \quad . \tag{3.2/5}$$

This equation can be solved with respect to $\underline{r}^k$ if and only if the matrix $(\underline{D}^k)^T\underline{P}\ \underline{D}^k$ is nonsingular. The matrix $(\underline{D}^k)^T\underline{P}\ \underline{D}^k$ has this property if and only if the column vectors of the matrix $\underline{D}^k$ are linearly independent because the matrix $\underline{P}$ is positive definite by virtue of the assumption made in chapter 2 (see appendix A). Assume that the matrix $(\underline{D}^k)^T\underline{P}\ \underline{D}^k$ is nonsingular the solution of eq. (3.2/5) is given by

$$\underline{r}^k = [(\underline{D}^k)^T\underline{P}\ \underline{D}^k]^{-1}(\underline{D}^k)^T\underline{P}\ \overline{\underline{e}}(\underline{x}^k) \quad . \tag{3.2/6}$$

For this vector $\underline{r}^k$ we really have a minimum of $\overset{\approx}{Q}$ because the matrix $(\underline{D}^k)^T\underline{P}\ \underline{D}^k$ - which is the Hessian matrix for $\overset{\approx}{Q}$ except for the factor 2 - is positive definite (see appendix A).

Then the next point $\underline{x}^{k+1}$ of the iteration - concerning the function $\overset{\approx}{Q}$ - is given by

$$\underline{x}^{k+1} = \underline{x}^k + [(\underline{D}^k)^T\underline{P}\ \underline{D}^k]^{-1}(\underline{D}^k)^T\underline{P}\ \overline{\underline{e}}(\underline{x}^k) \quad . \tag{3.2/7}$$

If we choose this point $\underline{x}^{k+1}$ as next point for the iteration concerning the function Q so it is evident that for this point $\underline{x}^{k+1}$ the inequality

$$Q(\underline{x}^{k+1}) < Q(\underline{x}^k) \tag{3.2/8}$$

has not to hold necessarily because the linear approximation (3.2/1) of $\underline{f}(\underline{x})$ is in general valid only in a small vicinity of $\underline{x}^k$. But in any case we demand the validity of the inequality (3.2/8) in order to guarantee that the sum of squares Q decreases in each stage of the iteration. This is a minimum requirement concerning the convergence of a minimizing method. The validity of the inequality (3.2/8) is called stability condition in the sequel. Methods which do not satisfy the stability condition (3.2/8) - that means methods for which it is not guaranteed that the sum of squares decreases in each stage of the iteration - are in general not suitable for the solution of the described problem. Therefore the iteration instruction (3.2/7) cannot usually be applied for the solution of nonlinear least squares problems. The method given by the instruction (3.2/7) is the so-called Gauss-Newton method [8, 9]. From it we obtain the well-known Newton method if we have m = n, that means if we have as many equations as unknown parameters, and if we choose the identity matrix $\underline{I}$ for the matrix $\underline{P}$. The iteration instruction for the Newton method is then given by

$$\underline{x}^{k+1} = \underline{x}^k + (\underline{D}^k)^{-1} \, \overline{\underline{e}}(\underline{x}^k) \tag{3.2/9}$$

where we have to assume that the Jacobian matrix $\underline{D}^k$ is nonsingular.

Returning to the above considered Gauss-Newton method we can write for eq. (3.2/7) because of eq. (3.1/13)

$$\underline{x}^{k+1} = \underline{x}^k - \frac{1}{2} [(\underline{D}^k)^T \underline{P} \, \underline{D}^k]^{-1} \underline{\nabla} Q(\underline{x}^k) \ . \tag{3.2/10}$$

This relation can be interpreted as an iteration instruction of an universal gradient method if we write

$$\underline{R}^k = \frac{1}{2} [(\underline{D}^k)^T \underline{P} \, \underline{D}^k]^{-1} \ . \tag{3.2/11}$$

So we have proved that the Gauss-Newton method may be considered as an universal gradient method when utilizing the matrix $\underline{R}^k$ given by eq. (3.2/11). Different from the universal gradient methods the step-length factor is determined a priori, namely $\alpha_{min}$ = 1. An obvious modification of the Gauss-Newton method is the determination of the step-length factor $\alpha_{min}$ by an one-dimensional search according to eq. (3.1/5). As the matrix $\underline{R}^k$ from eq. (3.2/11) is positive definite on the assumption that the column vectors of the matrix $\underline{D}^k$ are linearly independent and because of the assumed properties of the matrix $\underline{P}$ (see appendix A) the stability of the so modified Gauss-Newton method is guaranteed. This method was first proposed by Hartley [10]. Therefore we call it Hartley's method in the sequel.

The Gauss-Newton method as well as Hartley's method are only applicable if the column vectors of the matrix $\underline{D}^k$ are linearly independent in the course of the whole iteration. But this property cannot be guaranteed. It is always possible that the column vectors of the matrix $\underline{D}^k$ are linearly dependent in one or more stages of the iteration. Therefore we need a method which is also applicable if the column vectors of the matrix $\underline{D}^k$ are not always linearly independent.

Such a method was proposed by Levenberg [11]. Similar to the Gauss-Newton method, his method is also based on the linear approximation (3.2/1) of the function $\underline{f}(\underline{x})$. Levenberg does not consider the sum of squares $\overset{\approx}{Q}$ from eq. (3.2/3) but the modified sum of squares $\widetilde{Q}$ given by

$$\widetilde{Q} = \overset{\approx}{Q} + \lambda^k (\underline{r}^k)^T \underline{W}^k \underline{r}^k \ . \tag{3.2/12}$$

In eq. (3.2/12) the vector $\underline{r}^k$ is defined by eq. (3.2/2). $\lambda^k$ is a real non-negative scalar and $\underline{W}^k$ is a real positive-definite diagonal matrix. The adding of the term $(\underline{r}^k)^T \underline{W}^k \underline{r}^k$ is to cause the absolute value of the vector $\underline{r}^k$ not to increase very much when minimizing the function $\widetilde{Q}$. Therefore we have the hope that the vectors $\underline{x}^{k+1}$ and $\underline{x}^k$ do not differ very much so that $\underline{f}(\underline{x})$ is described sufficiently exactly by its linear approximation (3.2/1) so that the stability condition (3.2/8) is satisfied if we determine $\underline{x}^{k+1}$ for the iteration concerning Q by minimizing $\widetilde{Q}$.

The minimization of $\widetilde{Q}$ from eq. (3.2/12) leads to the relation

$$[\lambda^k \underline{W}^k + (\underline{D}^k)^T \underline{P} \, \underline{D}^k]\underline{r}^k = (\underline{D}^k)^T \underline{P} \, \underline{\bar{e}}(\underline{x}^k) \qquad (3.2/13)$$

for the direction vector $\underline{r}^k$.

Levenberg has shown that it is always possible to solve this system of linear equations by the choice of an appropriate scalar $\lambda^k \geq 0$ independent on the fact whether the column vectors of the matrix $\underline{D}^k$ are linearly independent or not. The solution of eq. (3.2/13) with respect to $\underline{r}^k$ is given by

$$\underline{r}^k = [\lambda^k \underline{W}^k + (\underline{D}^k)^T \underline{P} \, \underline{D}^k]^{-1} (\underline{D}^k)^T \underline{P} \, \underline{\bar{e}}(\underline{x}^k) \; . \qquad (3.2/14)$$

By the adding of the term $\lambda^k (\underline{r}^k)^T \underline{W} \, \underline{r}^k$ in eq. (3.2/12) it is at first guaranteed that the matrix on the left-hand side of eq. (3.2/13) is always invertible. This is easily shown if we bear in mind that the matrix $\underline{W}^k$ is positive definite and that the matrix $(\underline{D}^k)^T \underline{P} \, \underline{D}^k$ is at least positive semidefinite (see appendix A). Therefore the matrix $\lambda^k \underline{W}^k + (\underline{D}^k)^T \underline{P} \, \underline{D}^k$ is positive definite and consequently invertible. Thereby it is not guaranteed however that the stability condition (3.2/8) is satisfied. But Levenberg has shown that it is possible to satisfy the inequality (3.2/8) by the choice of a sufficiently great positive scalar $\lambda^k$.

Therefore this method represents one possibility for the solution of nonlinear least squares problems. But it is to remark that the choice of the scalar $\lambda^k$ is a problem which is not to undervalue. If we choose such a scalar $\lambda^k$ for which the stability condition (3.2/8) is not satisfied it is necessary to increase $\lambda^k$ and to solve the system of linear equations (3.2/13) once more. This processing requires a considerable effort.

This method developed by Levenberg was unknown for a long time. It was Marquardt who "rediscovered" this method [12]. Marquardt chose the identity matrix $\underline{I}$ for the matrix $\underline{W}^k$ of Levenberg's method.

Levenberg's method as well as Marquardt's method satisfy the iteration instruction

$$\underline{x}^{k+1} = \underline{x}^k + [\lambda^k \underline{W}^k + (\underline{D}^k)^T \underline{P} \, \underline{D}^k]^{-1}(\underline{D}^k)^T \underline{P} \, \underline{\bar{e}}(\underline{x}^k) \; . \qquad (3.2/15)$$

This iteration instruction can also be interpreted as an instruction of an universal gradient method if we set

$$\underline{R}^k = \frac{1}{2} [\lambda^k \underline{W}^k + (\underline{D}^k)^T \underline{P} \, \underline{D}^k]^{-1} \; . \qquad (3.2/16)$$

As in the Gauss-Newton method the step-length factor is determined a priori in Levenberg's and Marquardt's method. In contrast to the universal gradient method there is no determination of $\alpha_{min}$. This search procedure is substituted by the calculation of in general several matrices $\underline{R}^k$ in each stage of the iteration whereby the scalar $\lambda^k$ is increased until the stability condition (3.2/8) is satisfied.

By developing appropriate strategies for the modification of the scalar $\lambda^k$ it is possible to diminish the number of calculations of the matrices $\underline{R}^k$ in one stage of the iteration. One possible strategy for the selection of an appropriate value of $\lambda^k$ is given in [13]. Marquardt reports that it is necessary in general to calculate at least two matrices $\underline{R}^k$ in each stage of the iteration. One possible thought is to choose the scalar $\lambda^k$ always very great. But in this case Marquardt has shown that in the main we obtain a correction of $\underline{x}^k$ in the direction of the negative gradient, that means Marquardt's method turns over to the method of steepest descent with its well-known disadvantages.

This can be seen qualitatively by considering eq. (3.2/16). For a sufficiently great scalar $\lambda^k$ the matrix $(\underline{D}^k)^T\underline{P}\ \underline{D}^k$ can be neglected in comparison with $\lambda^k\underline{W}^k$ which is $\lambda^k\underline{I}$ in Marquardt's method. Therefore we obtain

$$\underline{R}^k \approx \frac{1}{2}\frac{1}{\lambda^k}\ \underline{I}\ . \tag{3.2/17}$$

The matrix $\underline{R}^k$ from eq. (3.2/17) yields the direction vector of the method of steepest descent except for the factor $(2\lambda^k)^{-1}$. From Levenberg's respectively Marquardt's method we obtain the Gauss-Newton method if we always set $\lambda^k = 0$ in eq. (3.2/16). The principal components of Marquardt's method are shown in the logic diagram 3.

A modification of Marquardt's method was given by R. R. Meyer [9]. Meyer has modified Marquardt's method in a similar way as Hartley did it for the Gauss-Newton method. In order to guarantee the validity of the stability condition (3.2/8) Meyer utilizes an one-dimensional search along the straight line $\underline{x}^k + \alpha\ \underline{r}^k$ whereby $\underline{r}^k$ is given by eq. (3.2/14). In Meyer's method the matrix $\underline{W}^k$ is given by

$$\underline{W}^k = \text{diag}\ ((\underline{D}^k)^T\underline{P}\ \underline{D}^k)\ . \tag{3.2/18}$$

$\underline{W}^k$ is that matrix which has on its main diagonal the elements of the main diagonal of the matrix $(\underline{D}^k)^T\underline{P}\ \underline{D}^k$ whereas all other elements are zero. In order to guarantee that this matrix $\underline{W}^k$ is positive definite Meyer has to assume that no column vector of the matrix $\underline{D}^k$ vanishes. But this is not always true (see for instance the example considered in subsection 7.2.4). Under the assumption that no column vector of the matrix $\underline{D}^k$ vanishes Meyer's method satisfies the stability condition (3.2/8). The advantage of Meyer's method in comparison with Levenberg's and Marquardt's method is that the system of linear equations (3.2/13) has to be solved only once in each stage of the iteration. As in Marquardt's method we also have the problem of modifying the scalar $\lambda^k$ from one stage of the iteration to another.

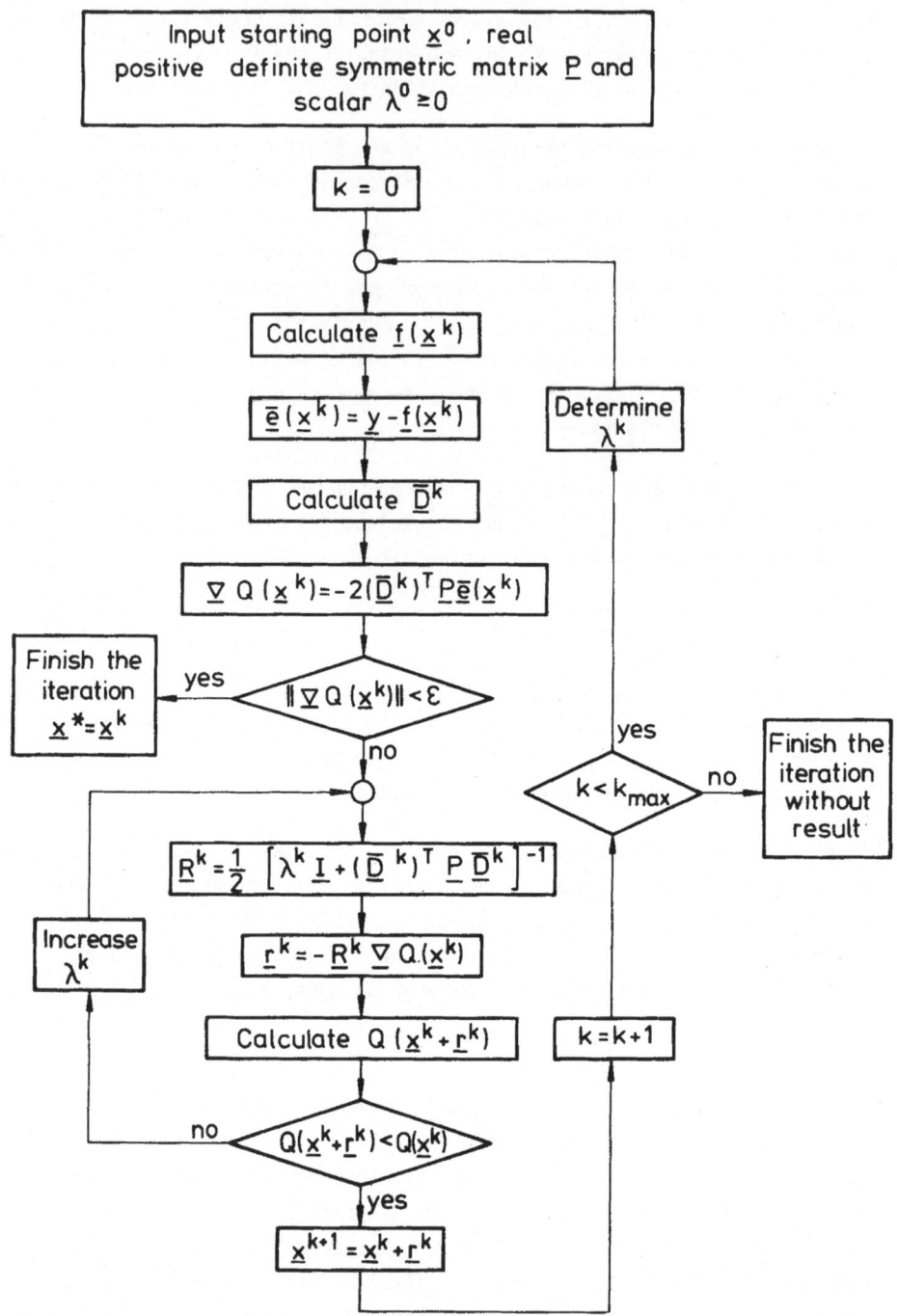

Logic diagram 3: Principal components of Marquardt's method for the solution of nonlinear least squares problems.

Meyer's method is an universal gradient method. Its principal components are shown in the logic diagram 4.

With these considerations we finish the survey of well-known special minimizing methods for the solution of nonlinear least squares problems. It is to remark that this survey is certainly not complete (for some further methods see for instance [1]).

### 3.3 Comparison of special minimizing methods with universal minimizing methods concerning the solution of nonlinear least squares problems

In the two preceding sections we have shown some possibilities how to solve nonlinear least squares problems with the help of universal and special minimizing methods. There is the obvious question which kind of method will be "better" suited for the solution of these problems. That means whether there is the chance of deriving a great benifit from the knowledge of the special structure (2/11) of the function to be minimized. Surely, the answer to this question cannot be given in a generally valid way because the problems which come into question differ very much. But before we try to answer some aspects of the asked question it is necessary to decide which criterion is to be used in order to judge whether a method is better than another. As mentioned at the beginning of section 3.1 there are many very different possible criteria. Assume we have chosen one criterion. Then we are not able to answer the asked question in a way which is always valid as long as we only consider some selected test problems. But this is the only practicable manner of obtaining results concerning minimizing methods on digital computers. It is clear that these results cannot be generalized arbitrarily but we obtain some qualitative indications to the performance of the methods.

It is also possible to judge the different minimizing methods by means of theoretical considerations concerning the way of convergence of the individual methods. These considerations are performed (see for instance [14]). The disadvantage of this processing is that we have to make a lot of assumptions referring to the function to be minimized. These assumptions are often not satisfied or cannot be examined in nonlinear least squares problems which are derived from practical applications. Often these assumptions lead to sufficient conditions for the convergence which are not necessary. Moreover problems can appear because of the limited accuracy of numerical calculations on digital computers. This fact is usually not taken into account in theoretical considerations. Therefore many of the results obtained by theoretical considerations are not very relevant to judge the performance of minimizing methods on digital computers. For these reasons selected test problems only come into question to judge the different minimizing methods.

Among others, Bard [15], McKeown [16], and Box [17] have made comparing investigations of minimizing methods for the solution of nonlinear least squares problems.

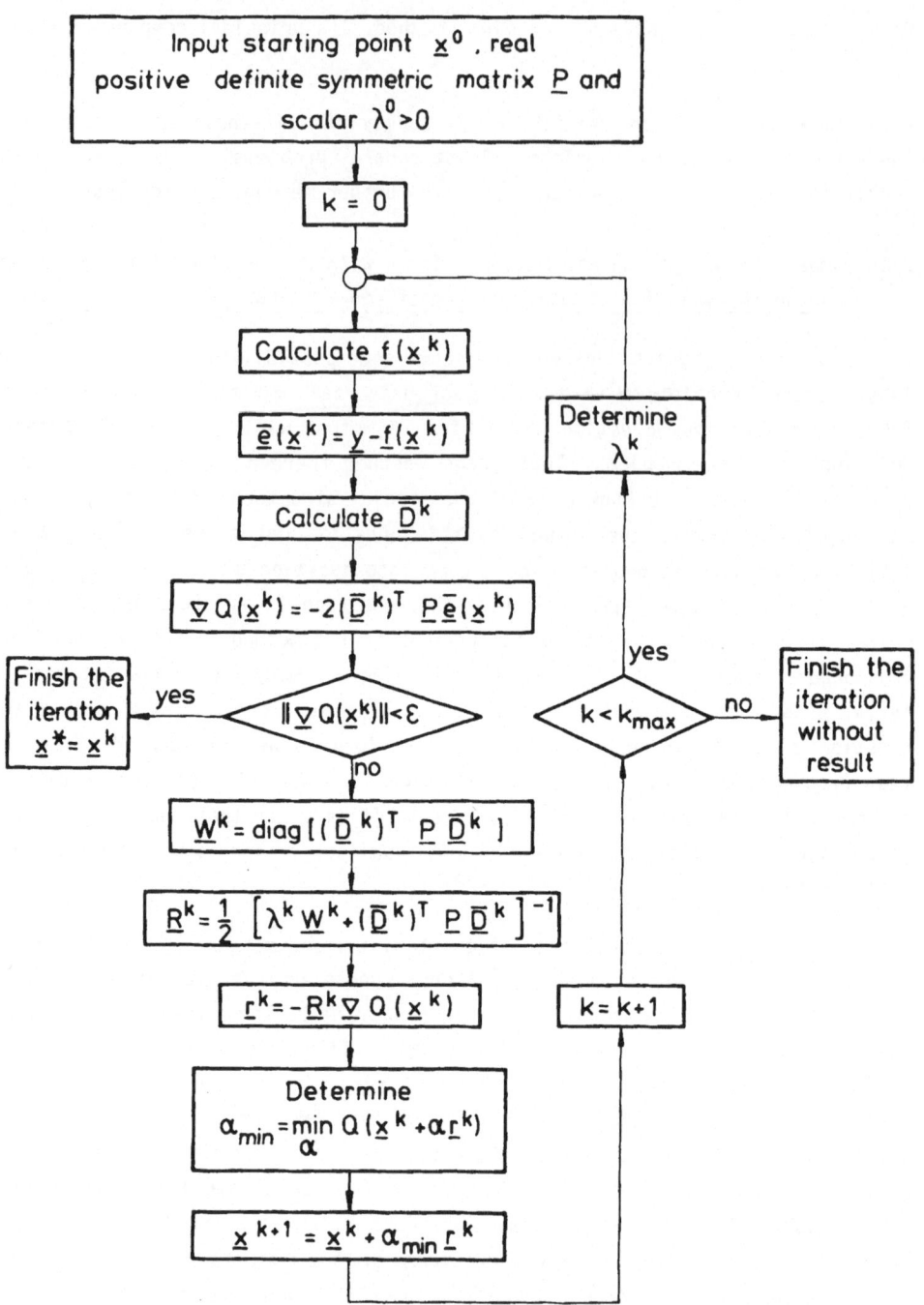

Logic diagram 4: Principal components of Meyer's method for the solution of nonlinear least squares problems.

Here we present a summary of the results which have been obtained by partly very extensive investigations.

Bard investigates five different minimizing methods. He utilizes these methods to-gether with different strategies for the determination of the step-length factor $\alpha_{min}$ Moreover he examines the Davidon-Fletcher-Powell method for five different bounds concerning the Euclidean length of the direction vector $\underline{r}^k$ and the difference of two succeeding parameter vectors $\underline{x}^k$ and $\underline{x}^{k+1}$. So altogether he obtains thirteen methods which are investigated. Four of these methods are special minimizing methods, namely the Gauss-Newton method, Marquardt's method and two versions of Hartley's method. The other nine methods are universal minimizing methods. Five of them are - as mention-ed - different versions of the Davidon-Fletcher-Powell method. All these methods are programmed in FORTRAN IV. The calculations are performed in single precision floating point arithmetic on an IBM System /360 Model 50.

The thirteen methods are performed with five test problems. One of these test pro-blems is utilized in six different versions so that we have a total of ten test pro-blems. In general it is possible to calculate the necessary derivatives of the func-tions $f_i(\underline{x})$ with respect to $x_j$, j = 1, ..., n from analytic formulas. In some cases where the model is given in the form of differential equations the derivatives are determined by numerical integration. The number of parameters to be determined in the test problems ranges from three to ten. The number of given fixed values $y_i$ ranges from eight to thirty. These values $y_i$ are artificially constructed; they are not measured really. Each test problem is run with all minimizing methods. After this all methods are classified for each test problem. The methods are divided into three classes:

Class 1: The results are practically indistinguishable from the best estimates ob-tained.

Class 2: The results are somewhat inferior to the best, but still acceptable.

Class 3: The results are unacceptable; the method does not converge.

The criteria utilized for this classification are not very rigid and it might be dif-ficult to perform this classification. But it is an easy task for Bard to classify each method uniquely in that class it belongs to.

Within classes the methods are listed in order of increasing numbers of equivalent function evaluations. The tool of equivalent function evaluations has been introduced by Box [17]. When determining the number of equivalent function evaluations we do not only count the number of function evaluations but also the number of derivative eva-luations. That means the evaluation of a gradient vector with n components is equi-valent to n function evaluations. Applying the rough classification together with this criterion we obtain a sequence of the performance of the methods for each test problem.

Bard's investigations show that the four special minimizing methods perform best in eight out of ten test problems. That means they are on the ranks one to four in the Class 1. For the two other test problems the special minimizing methods are also in Class 1 on front ranks. Hence it follows that the special minimizing methods are superior to the universal minimizing methods when considering the solution of nonlinear least squares problems. We can expect that this result is also valid for other problems than those examined by Bard. As to the sequence of the methods within the classes it is not possible to say that one method is really better than another because the differences are pretty small. To investigate this question it is necessary to consider more test problems.

It is to remark that there is a lot of problems not investigated by Bard. In this connexion the following questions are of great interest: What is the performance of the methods if we have more parameters to be determined? How do the methods perform if we have only finite differences for the calculation of the derivatives instead of analytic formulas? What is the performance of the methods if the given fixed values $y_i$ are not well fitted by the functions $f_i$? Are the sets of starting points $\underline{x}^0$ for which the methods converge very different and how is the performance for other starting points than the considered ones?

In the second above-mentioned investigation, McKeown examines six different minimizing methods. Four of them are special minimizing methods, among them Hartley's method. McKeown considers four different test problems. In these problems the derivatives of the functions $f_i(\underline{x})$ with respect to $x_j$, $j = 1, \ldots, n$ are not calculated by analytic formulas but by a numerical approximation. The given fixed values $y_i$ are really measured. The test problems are such with three to eight parameters $x_j$ and eight to twelve measured values $y_i$.

For judging the performance of the methods McKeown utilizes two performance criteria. These are the number of function evaluations and the computation time. The comparison by means of the computation time yields no serious results because some of McKeown's programs are written in FORTRAN IV and others in BASIC. Moreover he has no possibility to determine the computation time exactly to a certain extent. An error of $\pm$ 25 % in the computation time cannot be ruled out. Therefore the only reliable criterion is the number of function evaluations. Applying this criterion it turns out that the special minimizing methods are superior to the universal minimizing methods although the difference is not so striking as in the case of the test problems investigated by Bard. Among other things this is a consequence of the different test problems under consideration and of the different applied criteria. McKeown himself states that the number of function evaluations is not always an appropriate criterion for judging the performance of a minimizing method.

A further comparison of several minimizing methods for the solution of nonlinear least squares problems was carried out by Box [17]. Altogether Box considers eight minimizing methods. Six of them are universal minimizing methods and two are special minimizing methods. Only two of the six minimizing methods are universal gradient methods. The number of parameters $x_j$ ranges from two to twenty. The number of given fixed values $y_i$ ranges from five to twenty. Box considers five test problems. As criterion for the performance of the minimizing methods he utilizes the number of equivalent function evaluations. As conclusion of his investigations Box states that the special minimizing methods are to prefer to those methods which merely minimize the sum of squares.

From all these investigations it results that the special minimizing methods can be considered to be better appropriate for the solution of nonlinear least squares problems although the investigated minimizing methods and the criteria for judging their performance differ. Therefore it is intelligent to take advantage of the special structure of the function Q from eq. (2/11) when solving nonlinear least squares problems.

References:

[1]  Bard, Y.:
     Nonlinear Parameter Estimation
     New York and London: Academic Press (1974).
[2]  Himmelblau, D.M.:
     Applied Nonlinear Programming
     New York: McGraw-Hill, Inc. (1972).
[3]  Pierre. D.A.:
     Optimization Theory with Applications
     New York: J. Wiley & Sons, Inc. (1969).
[4]  v. Mangoldt, H.; Knopp, K.:
     Einführung in die Höhere Mathematik
     Zweiter Band, 14. Auflage
     Stuttgart: S. Hirzel Verlag (1974).
[5]  Wismer, D.A.; Chattergy, R.:
     Introduction to Nonlinear Optimization
     New York: Elsevier North-Holland, Inc. (1978).
[6]  Huang, H.Y.:
     Unified Approach to Quadratically Convergent Algorithms for Function Minimization
     J. Optimization Theory Appl. 6, 405-423 (1970).
[7]  Fletcher, R.; Powell, M.J.D.:
     A Rapidly Convergent Descent Method for Minimization
     The Computer Journal 6, 163-168 (1963).
[8]  Dennis, J.E.:
     Nonlinear Least Squares and Equations
     Proc. of the Conference on the State of the Art in Numerical Analysis, April 12th-15th, 1976. Edited by D. Jacobs.
     New York and London: Academic Press (1977).
[9]  Meyer, R.R.:
     Theoretical and Computational Aspects of Nonlinear Regression
     In "Nonlinear Programming" edited by J. B. Rosen, O. L. Mangasarian, K. Ritter
     New York and London: Academic Press (1970).

[10] Hartley, H.O.:
The Modified Gauss-Newton Method for the Fitting of Nonlinear Regression Functions by Least Squares
Technometrics 2, 269-280 (1961).
[11] Levenberg, K.:
A Method for the Solution of Certain Nonlinear Problems in Least Squares
Quart. Appl. Math. 2, 164-168 (1944).
[12] Marquardt, D.W.:
An Algorithm for Least-Squares Estimation of Nonlinear Parameters
SIAM J. Appl. Math. 2, 431-441 (1963).
[13] Shanno, D.F.:
Parameter Selection for Modified Newton Methods for Function Minimization
SIAM J. Numer. Anal. 3, 366-372 (1970).
[14] Ortega, J.M.; Rheinboldt, W.C.:
Iterative Solution of Nonlinear Equations in Several Variables
New York and London: Academic Press (1970).
[15] Bard, Y.:
Comparison of Gradient Methods for the Solution of Nonlinear Parameter Estimation Problems
SIAM J. Numer. Anal. 1, 157-186 (1970).
[16] McKeown, J.J.:
A Comparison of Methods for Solving Nonlinear Parameter Estimation Problems
Proc. 3rd I.F.A.C. Symposium on "Identification and Parameter Estimation", The Hague (1973).
[17] Box, M.J.:
A Comparison of Several Current Optimization Methods, and Use of Transformations in Constrained Problems
The Computer Journal 9, 67-77 (1966).

# 4. A NEW METHOD FOR THE SOLUTION OF NONLINEAR LEAST SQUARES PROBLEMS

## 4.1 Requirements for a method for the solution of nonlinear least squares problems

The above presented special minimizing methods for the solution of nonlinear least squares problems have serious disadvantages to some extent. The Gauss-Newton method and Hartley's method are only applicable if the column vectors of the Jacobian matrix $\underline{D}^k$ are linearly independent in the entire course of the iteration. This assumption represents an essential restriction of the applicability of these methods because the demanded property concerning the linear independence of the column vectors cannot be guaranteed a priori. Additionally, there is a further problem which emerges when applying these methods on digital computers.

If the column vectors of the matrix $\underline{D}^k$ are theoretically linearly independent that means when examining this property analytically, so it is sometimes very difficult to ascertain this property on digital computers. To it we must take into account that there are two principal sources for errors when applying a method on digital computers. On the one hand digital computers only possess a finite number of bits for the representation of a number. This finiteness naturally causes errors because we can only represent a finite subset of rational numbers and not the infinite set of all real numbers on digital computers. On the other hand, there are always rounding errors when performing arithmetic operations on digital computers. For these reasons there will often be differences between the theoretical rank of a matrix - that means that rank which is analytically determined - and the numerical rank - that means that rank which is numerically determined on digital computers.

For the determination of the rank of a matrix $\underline{D}^k$ on a digital computer it is usually necessary to introduce a bound $\varepsilon$. For instance, if a characteristic value becomes less than this bound $\varepsilon$ during the processing of a numerical method on a digital computer we have to consider the column vectors of the matrix $\underline{D}^k$ as linearly dependent. Depending on the absolute value of this bound $\varepsilon$ we can obtain different values for the rank of this matrix $\underline{D}^k$.

In connexion with this problem remember the well-known Gaussian elimination method for the solution of systems of linear equations [1]. The system of linear equations can be solved if no pivotal element vanishes in the course of the elimination, that means the absolute value of all pivotal elements has to be different from zero. When applying this method on digital computers, it is necessary to substitute the number zero by a small positive number $\varepsilon$, for instance $\varepsilon = 10^{-7}$ in single precision floating point arithmetic. Depending on this bound $\varepsilon$ the system of linear equations possesses a solution or none.

Returning to the problem of the determination of the rank of the matrix $\underline{D}^k$, it is clear that this rank is uniquely given when determining it analytically. But this is no longer true when determining it on digital computers. From this numerical point of view the rank of a matrix is no uniquely given number but there is a range of ranks.

In general this problem cannot be solved by using multiple precision arithmetic [2].

Therefore it is necessary in general to assign a rank to the matrix $\underline{D}^k$ in the course of the processing of a numerical method on digital computers. For this reason we have to demand that a practically useful method for the solution of nonlinear least squares problems is applicable and in particular stable if we have to assign a rank to the matrix $\underline{D}^k$ which can be different from the theoretical rank. Neither the Gauss-Newton method nor Hartley's method have this property because for their applicability it is necessary that the column vectors of the matrix $\underline{D}^k$ are linearly independent. This is different in the case of Marquardt's method. By the choice of a sufficiently great scalar $\lambda^k$ it is guaranteed that the matrix $\underline{R}^k$ from eq. (3.2/16) also exists when taking into account the problems arising from a numerical point of view. The disadvantage of Marquardt's method is the fact that it often happens that the system of linear equations (3.2/13) has to be solved several times in one stage of the iteration.

Considering Meyer's method, the problem of assigning a rank to the matrix $\underline{D}^k$ is not solved. In Meyer's method a value for $\lambda^k$ is fixed in each stage of the iteration. For this value it is not guaranteed that it is great enough to render possible the inversion of the matrix from eq. (3.2/16), when paying attention to numerical problems. If $\lambda^k$ is not great enough it will be necessary to return to Marquardt's method and that means to increase $\lambda^k$ adequately.

The problem of assigning a rank to the matrix $\underline{D}^k$ also results from another reason which is connected with the sensitivity of the components of the direction vector $\underline{r}^k$ due to disturbances in the system of linear equations from which the direction vector $\underline{r}^k$ is obtained for the different minimizing methods. This problem will be considered in details in section 4.5.

For these and the above-mentioned reasons we have to derive five requirements for an always applicable method for the solution of nonlinear least squares problems.

Requirement 1

The method has to be stable, that means:  It must be guaranteed that the stability condition (3.2/8) is satisfied in each stage of the iteration.

Requirement 2

The method has to be applicable also in the case that the column vectors of the Jacobian matrix $\underline{D}^k$ are not linearly independent.

Requirement 3

The method has to be stable if for numerical reasons we have to assign a rank to the matrix $\underline{D}^k$ which is not necessarily identical with the analytically determined rank of the matrix $\underline{D}^k$.

Requirement 4

The method shall be so conditioned that in each stage of the iteration a system of linear equations has to be solved only once for the determination of the direction vector $\underline{r}^k$.

Requirement 5

If the column vectors of the Jacobian matrix $\overline{\underline{D}}^k$ are linearly independent and if this property can be examined numerically the method shall turn over to Hartley's method.

We are to remark that the requirement 4 is essential in order to save computation time. The requirement 5 is important because of the convergence of the method. It is known that Hartley's method is able to determine the minimum of a quadratic function $Q(\underline{x})$ in one step. Therefore it can be expected that the convergence is good in the vicinity of the minimum even if the function to be minimized is not quadratic.

In the next sections a new numerical method is derived and discussed which satisfies the above-mentioned requirements.

## 4.2  Derivation of a new method for the solution of nonlinear least squares problems

To begin with we introduce some abbreviations in order to simplify the notation in the sequel. These follow from the decomposition of the matrix $\underline{P}$ given by eq. (A/2). By utilization of this decomposition we can write for the sum of squares $Q$ from eq. (2/11)

$$Q = \overline{\underline{e}}^T \underline{P} \, \overline{\underline{e}} = \overline{\underline{e}}^T \underline{S}^T \underline{S} \, \overline{\underline{e}} \quad . \tag{4.2/1}$$

Introducing the new error vector

$$\underline{e} = \underline{S} \, \overline{\underline{e}} \tag{4.2/2}$$

we obtain

$$Q = \underline{e}^T \underline{e} \quad . \tag{4.2/3}$$

instead of eq. (4.2/1).

With respect to the eqs. (A/2), (A/5), and (4.2/2) we can write for the gradient of the sum of squares $Q$ from eq. (3.1/13)

$$\begin{aligned} \underline{\nabla} Q(\underline{x}^k) &= - 2(\overline{\underline{D}}^k)^T \underline{P} \, \overline{\underline{e}}(\underline{x}^k) \\ &= - 2(\overline{\underline{D}}^k)^T \underline{S}^T \underline{S} \, \overline{\underline{e}}(\underline{x}^k) \\ &= - 2(\underline{D}^k)^T \underline{e}(\underline{x}^k) \quad . \end{aligned} \tag{4.2/4}$$

With these new notations we have made all preparations necessary for the derivation of a new method for the solution of nonlinear least squares problems. This new method is an universal gradient method as well as the methods presented in section 3.2. The structure of the methods from section 3.2 is shown in the logic diagram 2. Depending on the choice of the matrices $\underline{R}^k$ we obtain different minimizing methods. In the sequel we derive a new instruction for the determination of the matrix $\underline{R}^k$ in each stage of the iteration.

At first we demand that the direction vector $\underline{r}^k$ of the resulting minimizing method is given by

$$\underline{r}^k = \overline{\underline{R}}^k \underline{e}(\underline{x}^k) \tag{4.2/5}$$

whereby $\underline{e}(\underline{x}^k)$ is defined by eq. (4.2/2). Considering eq. (4.2/5) it cannot be recognized at this moment that the resulting minimizing method is an universal gradient method because the direction vector $\underline{r}^k$ is not the product of a matrix $\underline{R}^k$ and the gradient of $Q(\underline{x})$. This will be seen later on.

The principal components of the new method for the solution of nonlinear least squares problems are shown in the logic diagram 5. In comparison with the logic diagram 2, the decomposition of the real positive definite matrix $\underline{P}$ has come up. The matrix $\underline{P}$ is represented as the product of two triangular matrices which are transposed to each other. This decomposition is not problematic because its numerical execution is easy and quick [3, 4]. If $\underline{P}$ is a diagonal matrix - which often happens in technical applications - the decomposition is extremely simple. It is only necessary to compute the square roots of the diagonal elements of the matrix $\underline{P}$.

In the logic diagram 5 the determination of the matrix $\overline{\underline{R}}^k$ is not specified. The matrix $\overline{\underline{R}}^k$ has to be selected in such a way that the resulting minimizing method satisfies the requirements 1 to 5 from section 4.1. For that purpose we demand at first that the matrix $\overline{\underline{R}}^k$ satisfies the two equations

$$(\underline{D}^k \overline{\underline{R}}^k)^2 = \underline{D}^k \overline{\underline{R}}^k \tag{4.2/6}$$

and

$$(\underline{D}^k \overline{\underline{R}}^k)^T = \underline{D}^k \overline{\underline{R}}^k \ . \tag{4.2/7}$$

Eqs. (4.2/6) and (4.2/7) mean that $\overline{\underline{R}}^k$ has to be selected in such a way that $\underline{D}^k \overline{\underline{R}}^k$ is an orthogonal projector, i. e. an arbitrary vector $\underline{z} \in \mathrm{IR}^m$ is transformed into $\underline{z}''$ where $\underline{z}''$ is a vector in a yet unknown subspace $\mathbb{B} \subseteq \mathbb{R}^m$ and $\underline{z}^\perp = \underline{z} - \underline{z}''$ is a vector in the orthogonal complement $\mathbb{B}^\perp$ of $\mathbb{B}$ in $\mathbb{R}^m$.

Now we shall prove that the sum of squares Q does not increase in each stage of the iteration when determining the direction vector $\underline{r}^k$ by evaluating eq. (4.2/5) whereby the matrix $\overline{\underline{R}}^k$ is selected in such a way that it satisfies eqs. (4.2/6) and (4.2/7). In order to prove this it is sufficient to show that

$$(\underline{\nabla} Q(\underline{x}^k))^T \underline{r}^k \leq 0 \tag{4.2/8}$$

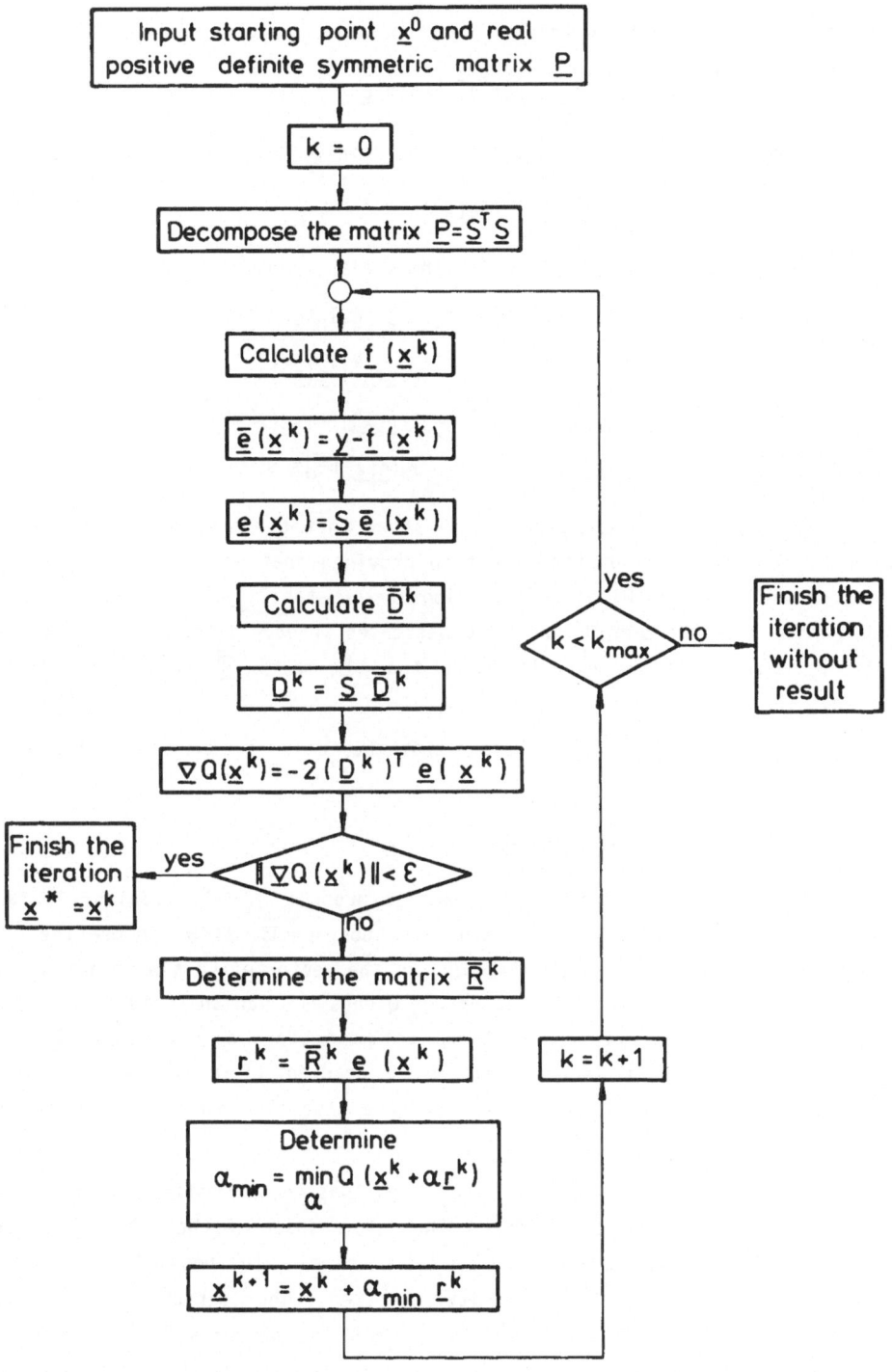

Logic diagram 5: Principal components of the new method for the solution of nonlinear least squares problems.

holds. In order to show this we decompose the vector $\underline{e}(\underline{x}^k) \in IR^m$ according to eq. (B/4)

$$\underline{e}(\underline{x}^k) = \underline{e}(\underline{x}^k)" + \underline{e}(\underline{x}^k)^{\perp} \text{ with } \underline{e}(\underline{x}^k)" \in \mathbb{B} \text{ and } \underline{e}(\underline{x}^k)^{\perp} \in \mathbb{B}^{\perp} \quad . \quad (4.2/9)$$

As the mapping $\underline{D}^k\underline{R}^k$ is an orthogonal projector on the subspace $\mathbb{B}$ the relation

$$\underline{D}^k\underline{R}^k\underline{e}(\underline{x}^k) = \underline{e}(\underline{x}^k)" \quad\quad\quad (4.2/10)$$

is valid. Therefore we can write for the scalar product (4.2/8)

$$\begin{aligned}
(\underline{\nabla} Q(\underline{x}^k))^T\underline{r}^k &= -2\,\underline{e}(\underline{x}^k)^T\underline{D}^k\underline{R}^k\underline{e}(\underline{x}^k) \\
&= -2\,\underline{e}(\underline{x}^k)^T\underline{e}(\underline{x}^k)" \\
&= -2[\underline{e}(\underline{x}^k)" + \underline{e}(\underline{x}^k)^{\perp}]^T\underline{e}(\underline{x}^k)" \\
&= -2\|\underline{e}(\underline{x}^k)"\|^2 \leq 0 \quad .
\end{aligned} \quad (4.2/11)$$

With the inequality (4.2/11) we have shown that the sum of squares Q does not increase in each stage of the iteration provided that we only choose the step-length factor small enough. Within the meaning of the stability condition (3.2/8) we want to guarantee that the sum of squares Q decreases in each stage of the iteration. In order to achieve this we have to demand that the matrix $\underline{R}^k$ satisfies a further condition.

By inspecting eq. (4.2/11) we state that the scalar product (4.2/8) vanishes if and only if

$$\underline{e}(\underline{x}^k)" = \underline{0} \quad\quad\quad (4.2/12)$$

holds. Eq. (4.2/12) means that the scalar product (4.2/11) vanishes if and only if the vector $\underline{e}(\underline{x}^k)$ has no component in the subspace $\mathbb{B}$. If we choose the column space $R(\underline{D}^k)$ of the matrix $\underline{D}^k$ as subspace $\mathbb{B}$, so the vector $\underline{e}(\underline{x}^k)$ always has a component in this space provided that the gradient of Q does not vanish. Only in this case the error vector $\underline{e}(\underline{x}^k)$ is orthogonal to all column vectors of the matrix $\underline{D}^k$ which can be seen by inspecting eq. (4.2/4). But if the gradient of Q vanishes the necessary condition for a minimum is satisfied. In this case the minimizing method is finished anyhow.

For this reason it is intelligent to demand that the orthogonal projector $\underline{D}^k\underline{R}^k$ transforms an arbitrary vector $\underline{z} \in \mathbb{R}^m$ in the column space of the matrix $\underline{D}^k$. Here we must remember the remarks made in section 4.1 concerning the possibilities of determining the rank of a given matrix on digital computers. When performing this task analytically it is possible to select a basis for the subspace $R(\underline{D}^k)$ out of the column vectors of the matrix $\underline{D}^k$. Often this cannot be done exactly on digital computers. Rather it is necessary to assign a certain rank 1 to the matrix $\underline{D}^k$ which is not necessarily identical with the rank r which is determined analytically. Therefore we select 1 linearly independent column vectors out of the n column vectors of the matrix $\underline{D}^k$. Without loss of generality we assume that these 1 vectors are the first 1 vectors $\underline{d}_1^k$, ..., $\underline{d}_1^k$. By an appropriate renumeration of the column vectors of the matrix $\underline{D}^k$ this can always be achieved. Therefore we can divide the mxn-matrix $\underline{D}^k$ into a mxl-matrix $\underline{D}_1^k$ and a mx(n-1)-matrix $\underline{D}_{n-1}^k$ according to

$$\underline{D}^k = (\underline{D}_1^k, \underline{D}_{n-1}^k) \quad . \tag{4.2/13}$$

In order to transform an arbitrary vector $\underline{z} \in IR^m$ orthogonally in the column space $R(\underline{D}_1^k)$ we have to demand that the matrix $\overline{R}^k$ satisfies the equation

$$\underline{D}^k \overline{R}^k \underline{D}_1^k = \underline{D}_1^k \tag{4.2/14}$$

besides the eqs. (4.2/6) and (4.2/7) (compare appendix B, in particular eq. (B/10)). Assume that the matrix $\overline{R}^k$ is chosen in such a way that it additionally satisfies eq. (4.2/14) then eq. (4.2/11) becomes

$$(\underline{\nabla} Q(\underline{x}^k))^T \underline{r}^k = - 2 \|\underline{e}(\underline{x}^k)"\|^2 \leq 0 \text{ with } \underline{e}(\underline{x}^k)" \in R(\underline{D}_1^k) \quad . \tag{4.2/15}$$

As the iteration has not yet been finished the gradient of Q does not vanish. Therefore there is at least one column vector of the matrix $\underline{D}^k$ which is not orthogonal to the error vector $\underline{e}(\underline{x}^k)$. If we select this vector - and eventually further linearly independent column vectors - in order to build up the matrix $\underline{D}_1^k$, so it is guaranteed that the vector $\underline{e}(\underline{x}^k)$ certainly has a component in the column space of the matrix $\underline{D}_1^k$. Therefore we have

$$\underline{e}(\underline{x}^k)" \neq \underline{0} \quad . \tag{4.2/16}$$

Hence it follows that the absolute value of $\underline{e}(\underline{x}^k)"$ is different from zero so that we obtain the inequality

$$(\underline{\nabla}Q(\underline{x}^k))^T \underline{r}^k < 0 \tag{4.2/17}$$

from eq. (4.2/15). The inequality (4.2/17) guarantees that the stability condition (3.2/8) is satisfied.

The satisfaction of the stability condition does not depend on the fact whether we can or cannot exactly determine the rank of the matrix $\underline{D}^k$ on a digital computer. In order to guarantee the stability of the resulting minimizing method it is sufficient to build up the matrix $\underline{D}_1^k$ with at least one vector $\underline{d}_j^k$ which is not perpendicular to the error vector $\underline{e}(\underline{x}^k)$. Under certain circumstances we can add further linearly independent vectors. But it is not necessary to utilize all linearly independent vectors. This statement is very important concerning the performance of the method on digital computers, because the question whether a vector is already linearly dependent or still linearly independent plays no part. Naturally the problem is also solved that certain column vectors of the matrix $\underline{D}^k$ are linearly dependent on others when examining this property analytically. These linearly dependent vectors are not utilized for the building up of the matrix $\underline{D}_1^k$. Nevertheless the method is stable.

Comprehensively we can state that the just proposed method satisfies the requirements 1 to 3 from section 4.1, if we determine the direction vector $\underline{r}^k$ according to eq. (4.2/5) whereby the matrix $\underline{R}^k$ has to satisfy the eqs. (4.2/6), (4.2/7), and (4.2/14) simultaneously.

Assume for the following considerations that we can determine the theoretical rank r of the matrix $\underline{D}^k$. Out of the column vectors of the matrix $\underline{D}^k$ we select r linearly independent vectors. With these vectors we build up the matrix $\underline{D}_r^k$. The other - linearly dependent - vectors are gathered up in the matrix $\underline{D}_{n-r}^k$. This matrix can be represented as

$$\underline{D}_{n-r}^k = \underline{D}_r^k \underline{U} \tag{4.2/18}$$

whereby the column vectors of the matrix $\underline{U}$ contain the coefficients of the linear combination which establish the relation between the linearly dependent and linearly independent column vectors of the matrix $\underline{D}^k$.

Applying the orthogonal projector $\underline{D}^k\underline{R}^k$ to the matrix $\underline{D}_{n-r}^k$ we obtain

$$\underline{D}^k\underline{R}^k\underline{D}_{n-r}^k = \underline{D}^k\underline{R}^k\underline{D}_r^k\underline{U} = \underline{D}_r^k\underline{U} = \underline{D}_{n-r}^k \ . \tag{4.2/19}$$

Building up the composed matrix $\underline{D}^k = (\underline{D}_r^k, \ \underline{D}_{n-r}^k)$ and applying the orthogonal projector $\underline{D}^k\underline{R}^k$ to it we have the relation

$$\underline{D}^k\underline{R}^k\underline{D}^k = \underline{D}^k \ . \tag{4.2/20}$$

Here we remark explicitly that the relation (4.2/20) is only valid if we are able to determine the theoretical rank r of the matrix $\underline{D}^k$. In this case which is not so interesting when considering the application of this method on digital computers we do not need the requirement (4.2/6) for the matrix $\underline{R}^k$. Rather it is sufficient to demand that the matrix $\underline{R}^k$ satisfies the eqs. (4.2/7) and (4.2/20) because the validity of eq. (4.2/6) is automatically given because of eq. (4.2/20).

Now we know the relations which the matrix $\underline{R}^k$ must satisfy in order to comply with the requirements 1 to 3 from section 4.1. So far the matrix $\underline{R}^k$ is not known explicitly respectively it is not known how to determine it. Now we specify a matrix $\underline{R}^k$ which satisfies the eqs. (4.2/6), (4.2/7), and (4.2/14).

For that purpose we start from the decomposition (4.2/13) of the matrix $\underline{D}^k$ whereby the column vectors of the matrix $\underline{D}_1^k$ are selected as described above. That means: Selection of at least one vector $\underline{d}_j^k$ which is not orthogonal to the error vector $\underline{e}(\underline{x}^k)$ and adding further linearly independent vectors under certain circumstances. Applying this selection procedure it is guaranteed that the column vectors of the matrix $\underline{D}_1^k$ are linearly independent. As mentioned in appendix A, the matrix $(\underline{D}_1^k)^T\underline{D}_1^k$ is invertible. Therefore the matrix

$$\underline{D}_1^{k+} = ((\underline{D}_1^k)^T\underline{D}_1^k)^{-1}(\underline{D}_1^k)^T \tag{4.2/21}$$

exists. This matrix $\underline{D}_1^{k+}$ is a 1xm-matrix which satisfies the following four equations (for the notation of the matrix $\underline{D}_1^{k+}$ see appendix D)

$$\underline{D}_1^k\underline{D}_1^{k+}\underline{D}_1^k = \underline{D}_1^k \tag{4.2/22}$$

$$(\underline{D}_1^k\underline{D}_1^{k+})^T = \underline{D}_1^k\underline{D}_1^{k+} \tag{4.2/23}$$

$$\underline{D}_1^{k+}\underline{D}_1^k\underline{D}_1^{k+} = \underline{D}_1^{k+} \tag{4.2/24}$$

$$(\underline{D}_1^{k+}\underline{D}_1^k)^T = \underline{D}_1^{k+}\underline{D}_1^k \quad . \tag{4.2/25}$$

The validity of these equations can easily be shown if we utilize the relation (4.2/21) for the matrix $\underline{D}_1^{k+}$ in the eqs. (4.2/22), (4.2/23), (4.2/24), and (4.2/25).

Now we set for the matrix $\underline{R}^k$

$$\underline{R}^k = \begin{pmatrix} \underline{D}_1^{k+} \\ \underline{C}^k \end{pmatrix} \tag{4.2/26}$$

whereby the matrix $\underline{C}^k$ has to satisfy the equation

$$\underline{D}_{n-1}^k \underline{C}^k = \underline{0} \quad . \tag{4.2/27}$$

$\underline{C}^k$ is a (n-1)xm-matrix.

We must show that the matrix $\underline{R}^k$ given by eq. (4.2/26) really satisfies the eqs. (4.2/6), (4.2/7), and (4.2/14).

From eq. (4.2/6) we obtain

$$(\underline{D}^k\underline{R}^k)^2 = \left( (\underline{D}_1^k, \underline{D}_{n-1}^k) \begin{pmatrix} \underline{D}_1^{k+} \\ \underline{C}^k \end{pmatrix} \right)^2$$

$$= (\underline{D}_1^k\underline{D}_1^{k+})^2 = \underline{D}_1^k\underline{D}_1^{k+} = \underline{D}^k\underline{R}^k \quad . \tag{4.2/28}$$

From eq. (4.2/7) we obtain

$$(\underline{D}^k\underline{R}^k)^T = (\underline{D}_1^k\underline{D}_1^{k+})^T = \underline{D}_1^k\underline{D}_1^{k+} = \underline{D}^k\underline{R}^k \quad . \tag{4.2/29}$$

From eq. (4.2/14) we obtain

$$\underline{D}^k\underline{R}^k\underline{D}_1^k = \underline{D}_1^k\underline{D}_1^{k+}\underline{D}_1^k = \underline{D}_1^k \quad . \tag{4.2/30}$$

So we have shown that the above specified matrix $\underline{R}^k$ satisfies the eqs. (4.2/6), (4.2/7), and (4.2/14). If we determine the direction vector $\underline{r}^k$ from eq. (4.2/5) by utilizing the matrix $\underline{\underline{R}}^k$ from eq. (4.2/26) the resulting minimizing method complies with the requirements 1 to 3 from section 4.1.

Another method which also complies with the requirements 1 to 3 is proposed in [2]. This method is very expensive because it is necessary to determine the eigenvalues of the matrix $(\underline{\underline{D}}^k)^T\underline{\underline{D}}^k$ and the corresponding eigenvectors. Moreover the requirement 4 from section 4.1 is not satisfied.

By utilization of eq. (4.2/26) we can write the direction vector $\underline{r}^k$ as

$$\underline{r}^k = \begin{pmatrix} \underline{\underline{D}}_1^{k+} \\ \underline{\underline{C}}^k \end{pmatrix} \underline{e}(\underline{x}^k) \quad . \tag{4.2/31}$$

As we recognize, the first l components of the direction vector $\underline{r}^k$ are determined by the selection of l linearly independent column vectors $\underline{d}_1^k, \ldots, \underline{d}_l^k$. The other n-l components of the vector $\underline{r}^k$ can be influenced by the matrix $\underline{\underline{C}}^k$. This matrix $\underline{\underline{C}}^k$ has to satisfy eq. (4.2/27). One possible choice for the matrix $\underline{\underline{C}}^k$ is

$$\underline{\underline{C}}^k = \underline{\underline{0}} \quad . \tag{4.2/32}$$

Naturally this choice for the matrix $\underline{\underline{C}}^k$ guarantees the satisfaction of eq. (4.2/27). The general solution of eq. (4.2/27) with respect to the matrix $\underline{\underline{C}}^k$ is discussed in appendix C. The matrix $\underline{\underline{R}}^k$ from eq. (4.2/26) with $\underline{\underline{C}}^k = \underline{\underline{0}}$ satisfies the equation $\underline{\underline{R}}^k\underline{\underline{D}}^k\underline{\underline{R}}^k = \underline{\underline{D}}^k$ besides the eqs. (4.2/6), (4.2/7), and (4.2/14). This can easily be shown by evaluating this relation (compare appendix D).

In the sequel we always utilize the solution (4.2/32) for the matrix $\underline{\underline{C}}^k$. With this choice we obtain for the direction vector

$$\underline{r}^k = \underline{\underline{R}}^k\underline{e}(\underline{x}^k) = \begin{pmatrix} \underline{\underline{D}}_1^{k+} \\ \underline{\underline{0}} \end{pmatrix} \underline{e}(\underline{x}^k) \quad . \tag{4.2/33}$$

By inspecting eq. (4.2/33) we recognize that only the first l components of the direction vector $\underline{r}^k$ have to be determined. The other n-l components are fixed to zero. The first l components of the direction vector $\underline{r}^k$ are gathered up in the vector $\underline{r}_1^k$. Because of the relation (4.2/33) this vector $\underline{r}_1^k$ is given by

$$\underline{r}_1^k = \underline{\underline{D}}_1^{k+}\underline{e}(\underline{x}^k) = ((\underline{\underline{D}}_1^k)^T\underline{\underline{D}}_1^k)^{-1}(\underline{\underline{D}}_1^k)^T\underline{e}(\underline{x}^k) \quad . \tag{4.2/34}$$

The last equation can be written as

$$(\underline{\underline{D}}_1^k)^T\underline{\underline{D}}_1^k\underline{r}_1^k = (\underline{\underline{D}}_1^k)^T\underline{e}(\underline{x}^k) \quad . \tag{4.2/35}$$

Eq. (4.2/35) represents a system of linear equations for the determination of the components of the vector $\underline{r}_1^k$. This system of linear equations has to be solved only once in each stage of the iteration. This is the requirement 4 from section 4.1 for a method for the solution of nonlinear least squares problems. The meaning of the solution $\underline{r}_1^k$ from eq. (4.2/34) is discussed in details in appendix D.

It is left to show that the proposed method also complies with the requirement 5 from section 4.1. For that purpose we must take into account that Hartley's method is only applicable if the column vectors of the matrix $\underline{D}^k$ are linearly independent in each stage of the iteration. Assume that the matrix $\underline{D}^k$ has this property. That means we have n linearly independent column vectors in each stage of the iteration. Setting 1 = n in eq. (4.2/33) we obtain for the direction vector $\underline{r}^k$

$$\underline{r}^k = \underline{D}^{k+}\underline{e}(\underline{x}^k) = ((\underline{D}^k)^T\underline{D}^k)^{-1}(\underline{D}^k)^T\underline{e}(\underline{x}^k) \ . \tag{4.2/36}$$

But this is just the direction vector $\underline{r}^k$ of the Gauss-Newton method respectively Hartley's method if we pay attention to the relations between $\overline{\underline{D}}^k$ and $\underline{D}^k$ respectively between $\overline{\underline{e}}(\underline{x}^k)$ and $\underline{e}(\underline{x}^k)$ as they are given by eqs. (A/5), (4.2/2), and (A/2). So it has been shown that the proposed method yields the approved Hartley's method if the column vectors of the matrix $\underline{D}^k$ are always linearly independent and if we are able to determine this property on a digital computer.

So we have substantiated that the proposed method complies with all five requirements from section 4.1 for a method for the solution of nonlinear least squares problems.

Now we shall show that the just derived new method for the solution of nonlinear least squares problems is an universal gradient method. For that purpose we write the direction vector $\underline{r}^k$ from eq. (4.2/33) in a slightly modified way

$$\underline{r}^k = \begin{pmatrix} \underline{D}_1^{k+} \\ \underline{0} \end{pmatrix} \underline{e}(\underline{x}^k) = \begin{pmatrix} ((\underline{D}_1^k)^T\underline{D}_1^k)^{-1}(\underline{D}_1^k)^T \\ \underline{0} \end{pmatrix} \underline{e}(\underline{x}^k)$$

$$= \begin{pmatrix} ((\underline{D}_1^k)^T\underline{D}_1^k)^{-1} & \underline{0} \\ & \underline{0} \end{pmatrix} (\underline{D}^k)^T\underline{e}(\underline{x}^k)$$

$$= -\frac{1}{2} \begin{pmatrix} ((\underline{D}_1^k)^T\underline{D}_1^k)^{-1} & \underline{0} \\ & \underline{0} \end{pmatrix} \underline{\nabla} Q(\underline{x}^k) \ . \tag{4.2/37}$$

Inspecting the last line we recognize that the proposed method is an universal gradient method for we can write the direction vector $\underline{r}^k$ as

$$\underline{r}^k = -\underline{R}^k \underline{\nabla} Q(\underline{x}^k) \tag{4.2/38}$$

with

$$\underline{R}^k = \frac{1}{2} \begin{pmatrix} ((\underline{D}_1^k)^T\underline{D}_1^k)^{-1} & \underline{0} \\ & \underline{0} \end{pmatrix} \ . \tag{4.2/39}$$

The proposed new method follows from the logic diagram 2 where the principal components of an universal gradient method for the solution of nonlinear least squares problems are represented. Additionally we have to determine the matrix $\underline{R}^k$ from eq. (4.2/39) by the selection of 1 linearly independent column vectors $\underline{d}_1^k, \ldots, \underline{d}_l^k$ for the building up of the matrix $\underline{D}_{-l}^k$. The principal components of the new method together with the modifications introduced in the logic diagram 5 are represented in the logic diagram 6.

Until now, we have not considered the strategy for the selection of the column vectors of the matrix $\underline{D}^k$ for the building up of the matrix $\underline{D}_l^k$. Possible strategies for the solution of this problem are discussed in details in the sections 4.3 and 4.4.

But previously we want to interpret the solution (4.2/33) for the direction vector $\underline{r}^k$. This solution means that under certain circumstances some components of the parameter vector $\underline{x}^k$ are not modified in the stage k of the iteration. This happens if the number of the selected linearly independent column vectors is less than the number n of the column vectors of the matrix $\underline{D}^k$. It can be intelligent to keep some parameters constant if we consider the meaning of the column vectors of the matrix $\underline{D}^k$. For one column vector $\underline{d}_j^k$, $1 \leq j \leq n$, the relation

$$\underline{d}_j^k = \underline{d}_j(\underline{x}^k) = \underline{S}\,\overline{\underline{d}}_j(\underline{x}^k) = \underline{S}\left.\begin{pmatrix} \dfrac{\partial f_1}{\partial x_j} \\ \vdots \\ \dfrac{\partial f_m}{\partial x_j} \end{pmatrix}\right|_{\underline{x}^k} \tag{4.2/40}$$

holds because of the eqs. (A/5) and (3.1/11). Eq. (4.2/40) tells us that a column vector $\underline{d}_j^k$ describes the variations of the functions $f_1, \ldots, f_m$ with respect to the parameter $x_j$ at the point $\underline{x}^k$ if we neglect the unimportant multiplication with the matrix $\underline{S}$. If the column vectors of the matrix $\underline{D}^k$ are linearly dependent this means that at least one column vector $\underline{d}_j^k$ can be represented by a linear combination of the linearly independent column vectors. This means that the variations of the functions $f_1, \ldots, f_m$ with respect to the parameter $x_j$ can also be achieved by appropriate variations of the functions $f_1, \ldots, f_m$ with respect to the other parameters. Therefore we cannot attain a certain variation uniquely. Hence it is an obvious alternative to keep certain parameters constant because the variations of the functions $f_1, \ldots, f_m$ with respect to these parameters can also be realized by the variations caused by the other parameters. Just this is the way the proposed method proceeds.

If the column vectors of the matrix $\underline{D}^k$ are not really linearly dependent but only nearly linearly dependent the same argumentation as above is valid.

If the column vectors of the matrix $\underline{D}^k$ are linearly dependent there is the further possibility of modifying all parameters a little bit. This is done by a method which is often proposed in order to overcome the difficulties with Hartley's method [5].

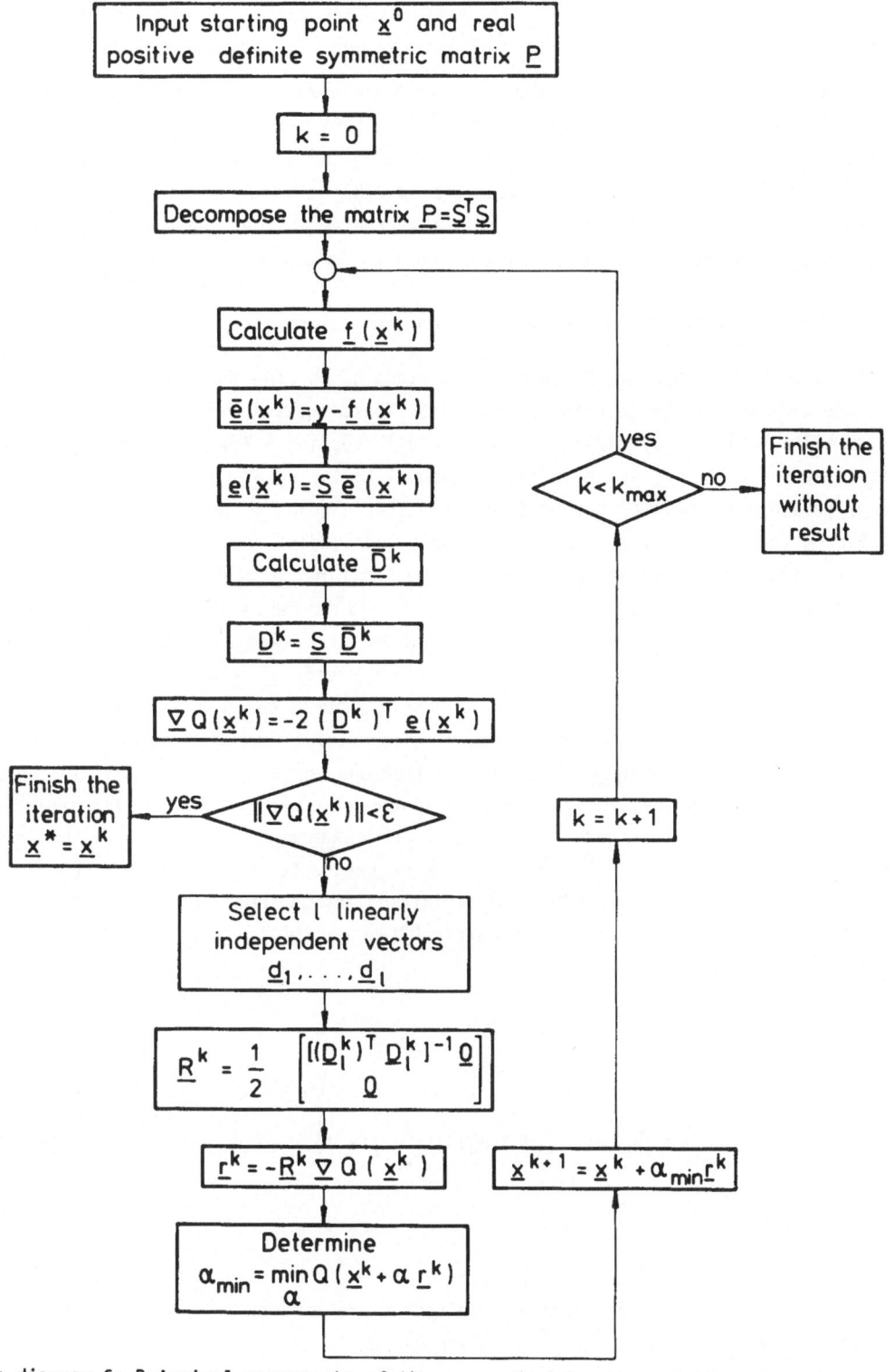

Logic diagram 6: Principal components of the new method for the solution of nonlinear least squares problems with determination of the direction vector $\underline{r}^k$.

Taking numerical problems into account this method can be problematic referring to the stability when applying it on digital computers. For details see appendix E.

In [6] another method is described for which under certain circumstances some components of the parameter vector are not modified in each stage of the iteration. In this method the value of the pivotal elements in the Gauss-Jordan method for the solution of the system of linear equations (3.2/5) is used to determine which components are to be modified and which not. The problem whether the resulting method for the solution of nonlinear least squares problems is always stable is not treated in [6].

## 4.3 The problem of the selection of the column vectors for the matrix $\underline{D}_1^k$

Apart from the selection of the column vectors $\underline{d}_j^k$ for the building up of the matrix $\underline{D}_1^k$ the method for the solution of nonlinear least squares problems proposed in section 4.2 is uniquely specified. As mentioned in section 4.2 it is necessary to select at least one vector $\underline{d}_j^k$ which is not orthogonal to the error vector $\underline{e}(\underline{x}^k)$. This is always possible as long as the gradient of Q at the point $\underline{x}^k$ does not vanish. To this selected vector we can add further linearly independent vectors. Now we have to put the question which vector $\underline{d}_j^k$ shall be selected first if there are several vectors which are not orthogonal to the error vector $\underline{e}(\underline{x}^k)$. Certainly it is intelligent to determine the vector $\underline{d}_j^k$ to be selected first in such a way that the slope of the sum of squares Q - which results for the direction vector $\underline{r}^k$ - at the point $\underline{x}^k$ becomes minimal if we only select one vector $\underline{d}_j^k$ for the building up of the matrix $\underline{D}_1^k$. This means we produce that direction vector $\underline{r}^k$ which yields a maximal decrease of the sum of squares Q at the point $\underline{x}^k$ if we only select one vector $\underline{d}_j^k$ altogether. But this is the situation at the beginning of the selection procedure because the linear independence of the column vectors has not yet been examined. The slope of the sum of squares Q at the point $\underline{x}^k$ in the direction of the vector $\underline{r}^k$ is given by $(\nabla Q(\underline{x}^k))^T \underline{r}^k$ if $\underline{r}^k$ is the direction vector for the minimizing method. When utilizing only one vector $\underline{d}_j^k$ for the building up of the matrix $\underline{D}_1^k$ we obtain for the direction vector $\underline{r}^k$ from eq. (4.2/33)

$$
\underline{r}^k = \begin{pmatrix} 0 \\ \vdots \\ 0 \\ ((\underline{d}_j^k)^T \underline{d}_j^k)^{-1} (\underline{d}_j^k)^T \underline{e}(\underline{x}^k) \\ 0 \\ \vdots \\ 0 \end{pmatrix}. \tag{4.3/1}
$$

We recognize that all components of the vector are zero except for the component j. So the scalar product $(\nabla Q(\underline{x}^k))^T \underline{r}^k$ is given by

$$(\underline{\nabla}Q(\underline{x}^k)^T\underline{r}^k = - 2 \ \underline{e}(\underline{x}^k)^T\underline{d}_j^k((\underline{d}_j^k)^T\underline{d}_j^k)^{-1}(\underline{d}_j^k)^T\underline{e}(\underline{x}^k) \ . \qquad (4.3/2)$$

The expression (4.3/2) has to be calculated for all vectors $\underline{d}_1^k$, ..., $\underline{d}_n^k$. That vector $\underline{d}_j^k$ for which the scalar product (4.3/2) becomes minimal is selected first. It can be shown (compare eq. (4.6/5)) that this is that vector $\underline{d}_j^k$ which is "least of all" orthogonal to the error vector $\underline{e}(\underline{x}^k)$. That means it is that vector $\underline{d}_j^k$ for which the absolute value of the angle between the vectors $\underline{d}_j^k$ and $\underline{e}(\underline{x}^k)$ is minimal. The vector $\underline{d}_j^k$ determined in this way is the first vector for the building up of the matrix $\underline{D}_l^k$. By an appropriate renumeration we can always achieve that this is the vector $\underline{d}_1^k$. As a matter of principle we need not select further vectors for the building up of the matrix $\underline{D}_l^k$ because the method is also stable if we only utilize one vector. Certainly this processing is not always intelligent if we remember the Gauss-Newton method or Hartley's method where all components of the parameter vector are modified in general. Modifying only one component of the parameter vector in each stage of the iteration usually causes a slow convergence. Especially this is true if there is a strong interaction between the components of the parameter vector to be determined.

Therefore we select more vectors $\underline{d}_j^k$ for the building up of the matrix $\underline{D}_l^k$. With the exception of the restriction that the vectors to be selected have to be linearly independent the way of selecting them is not prescribed.

The following processing is obvious: Select that vector $\underline{d}_j^k$ for which the scalar product (4.3/2) becomes minimal. Then examine the other vectors whether they are linearly dependent on the just selected vector. The vectors which are linearly dependent or "nearly" linearly dependent are eliminated. From the linearly independent vectors select one further vector. For this selection several strategies can be utilized which are discussed in details in section 4.4. Then examine the linear dependence of the remaining vectors on the two selected vectors. From the linearly independent vectors select a further one and so on. Apply this procedure until a subset of l linearly independent vectors is found which are utilized for the building up of the matrix $\underline{D}_l^k$. For this matrix $\underline{D}_l^k$ the matrix $\underline{D}_l^{k+}$ from eq. (4.2/21) is calculated. From $\underline{D}_l^{k+}$ the direction vector $\underline{r}^k$ is obtained by applying eq. (4.2/33).

To perform this selection procedure it is necessary to decide whether a vector under consideration is linearly dependent on the already selected vectors or not. In the sequel we describe a method which is able to examine the property of the linear independence and which simultaneously yields the inverse matrix $((\underline{D}_l^k)^T\underline{D}_l^k)^{-1}$ belonging to the already selected vectors $\underline{d}_1^k$, ..., $\underline{d}_l^k$. This inverse matrix is needed for the determination of the direction vector $\underline{r}^k$. For the determination of this inverse matrix we utilize values in the main which have to be calculated for the examination of the linear independence of a vector $\underline{d}_j^k$ in any case. So it is not necessary to compute the inverse matrix after having selected the subset of linearly independent vectors. This is convenient when looking for as short a computation time as possible.

In the sequel we describe a method for the examination of the linear independence and for the simultaneous calculation of the matrix $((\underline{D}_1^k)^T\underline{D}_1^k)^{-1}$. This method is based on Frobenius's relation for inversion [3, 7]. For the application of this relation it is necessary to partition a given quadratic nxn-matrix $\underline{A}$ in the way

$$\underline{A} = \begin{pmatrix} \underline{A}_1 & \underline{A}_2 \\ \underline{A}_3 & \underline{A}_4 \end{pmatrix} \quad . \tag{4.3/3}$$

The matrices $\underline{A}_1$ and $\underline{A}_4$ are quadratic. Therefore the matrices $\underline{A}_2$ and $\underline{A}_3$ are in general rectangular. The following theorem is valid [8]:

If both the matrix $\underline{A}_1$ and the matrix

$$\underline{A}_0 = \underline{A}_4 - \underline{A}_3\underline{A}_1^{-1}\underline{A}_2 \tag{4.3/4}$$

are nonsingular, the inverse matrix $\underline{A}^{-1}$ of $\underline{A}$ exists and it is given by

$$\underline{A}^{-1} = \begin{pmatrix} \underline{B}_1 & \underline{B}_2 \\ \underline{B}_3 & \underline{B}_4 \end{pmatrix} \tag{4.3/5}$$

whereby the matrices $\underline{B}_i$ are given by

$$\begin{aligned} \underline{B}_1 &= \underline{A}_1^{-1} + \underline{A}_1^{-1}\underline{A}_2\underline{A}_0^{-1}\underline{A}_3\underline{A}_1^{-1} \\ \underline{B}_2 &= -\underline{A}_1^{-1}\underline{A}_2\underline{A}_0^{-1} \\ \underline{B}_3 &= -\underline{A}_0^{-1}\underline{A}_3\underline{A}_1^{-1} \\ \underline{B}_4 &= \underline{A}_0^{-1} \quad . \end{aligned} \tag{4.3/6}$$

The validity of the given relations can easily be shown by evaluating the product of the matrices $\underline{A}$ and $\underline{A}^{-1}$.

The application of the just mentioned theorem for the problem under consideration - namely the examination of the linear independence of the vectors $\underline{d}_1^k$, ..., $\underline{d}_n^k$ and the calculation of the matrix $((\underline{D}_1^k)^T\underline{D}_1^k)^{-1}$ - stems from the following reflection:
Assume we have selected 1 linearly independent vectors $\underline{d}_1^k$, ..., $\underline{d}_1^k$. With these vectors we build up the matrix

$$\underline{D}_1^k = (\underline{d}_1^k, \ldots, \underline{d}_1^k) \quad . \tag{4.3/7}$$

Now we select a further vector $\underline{d}_{1+1}^k$ and add it to the already selected vectors. With these 1 + 1 vectors we build up the matrix

$$\underline{D}_{1+1}^k = (\underline{d}_1^k, \ldots, \underline{d}_1^k, \underline{d}_{1+1}^k) = (\underline{D}_1^k, \underline{d}_{1+1}^k) \quad . \tag{4.3/8}$$

Now we have to decide whether the vector $\underline{d}_{1+1}^k$ is linearly independent on the already selected vectors $\underline{d}_1^k$, ..., $\underline{d}_1^k$ or not. For this decision we make use of the following theorem [8]:

The vector $\underline{d}_{l+1}^k$ is linearly independent on the already selected vectors $\underline{d}_1^k, \ldots, \underline{d}_l^k$ if and only if

$$\underline{d}_{l+1}^{k\perp} \neq \underline{0} \text{ with } \underline{d}_{l+1}^{k\perp} \in R(\underline{D}_l^k)^\perp \tag{4.3/9}$$

is valid.

In order to apply this theorem it is necessary to calculate the vector $\underline{d}_{l+1}^{k\perp}$ and to examine whether it is different from the null vector. Naturally, the question concerning the linear independence of the vector $\underline{d}_{l+1}^k$ can also be answered by considering the value of $\|\underline{d}_{l+1}^{k\perp}\|$. If the vector $\underline{d}_{l+1}^k$ is linearly independent on the already selected vectors $\underline{d}_1^k, \ldots, \underline{d}_l^k$ then the matrix $((\underline{D}_{l+1}^k)^T\underline{D}_{l+1}^k)^{-1}$ exists (see appendix A). This matrix is needed for the determination of the direction vector $\underline{r}^k$. Now we show how it is possible to clear up the question of the linear independence and to calculate the matrix $((\underline{D}_{l+1}^k)^T\underline{D}_{l+1}^k)^{-1}$ simultaneously on the basis of the matrix $((\underline{D}_l^k)^T\underline{D}_l^k)^{-1}$ if the vector $\underline{d}_{l+1}^k$ is linearly independent on the vectors $\underline{d}_1^k, \ldots, \underline{d}_l^k$.

For that purpose we consider the matrix

$$(\underline{D}_{l+1}^k)^T\underline{D}_{l+1}^k = \begin{pmatrix} (\underline{D}_l^k)^T \\ (\underline{d}_{l+1}^k)^T \end{pmatrix} (\underline{D}_l^k, \underline{d}_{l+1}^k)$$

$$= \begin{pmatrix} (\underline{D}_l^k)^T\underline{D}_l^k & (\underline{D}_l^k)^T\underline{d}_{l+1}^k \\ (\underline{d}_{l+1}^k)^T\underline{D}_l^k & (\underline{d}_{l+1}^k)^T\underline{d}_{l+1}^k \end{pmatrix} . \tag{4.3/10}$$

Comparing the relation (4.3/10) with eq. (4.3/3) we find the following equivalents

$$\underline{A} = (\underline{D}_{l+1}^k)^T\underline{D}_{l+1}^k$$

$$\underline{A}_1 = (\underline{D}_l^k)^T\underline{D}_l^k$$

$$\underline{A}_2 = (\underline{D}_l^k)^T\underline{d}_{l+1}^k$$

$$\underline{A}_3 = (\underline{d}_{l+1}^k)^T\underline{D}_l^k = \underline{A}_2^T$$

$$\underline{A}_4 = (\underline{d}_{l+1}^k)^T\underline{d}_{l+1}^k . \tag{4.3/11}$$

In order to apply the above-mentioned theorem it is necessary that the matrices $(\underline{D}_l^k)^T\underline{D}_l^k$ and

$$\underline{A}_0 = (\underline{d}_{l+1}^k)^T\underline{d}_{l+1}^k - (\underline{d}_{l+1}^k)^T\underline{D}_l^k((\underline{D}_l^k)^T\underline{D}_l^k)^{-1}(\underline{D}_l^k)^T\underline{d}_{l+1}^k \tag{4.3/12}$$

are nonsingular. By virtue of the assumption the matrix $(\underline{D}_l^k)^T\underline{D}_l^k$ is nonsingular. It remains to examine under which conditions the matrix $\underline{A}_0$ from eq. (4.3/12) is nonsingular. By inspecting eq. (4.3/12) we state that the matrix $\underline{A}_0$ is a 1x1-matrix, that means it is a scalar. This scalar shall be denoted by $a_l^k$. The index 1 is chosen in order to remember that the scalar $a_1^k$ refers to 1 selected vectors $\underline{d}_1^k, \ldots, \underline{d}_1^k$. To guarantee the existence of the reciproke of $a_1^k$ the scalar $a_1^k$ has to be different from zero. The scalar $a_1^k$ is given by

$$a_1^k = (\underline{d}_{1+1}^k)^T [\underline{I} - \underline{D}_1^k ((\underline{D}_1^k)^T \underline{D}_1^k)^{-1} (\underline{D}_1^k)^T] \underline{d}_{1+1}^k$$

$$= (\underline{d}_{1+1}^k)^T [\underline{I} - \underline{D}_1^k \underline{D}_1^{k+}] \underline{d}_{1+1}^k \quad . \tag{4.3/13}$$

By inspecting eqs. (4.2/22) and (4.2/23) we recognize that the matrix $\underline{D}_1^k \underline{D}_1^{k+}$ is an orthogonal projector because it is idempotent and symmetric. Because of eq. (4.2/22) this orthogonal projector takes the vector $\underline{d}_{1+1}^k$ into the column space of the matrix $\underline{D}_1^k$.

Decompose the vector $\underline{d}_1^{k+1}$ according to

$$\underline{d}_{1+1}^k = \underline{d}_{1+1}^{k''} + \underline{d}_{1+1}^{k\perp} \text{ with } \underline{d}_{1+1}^{k''} \in R(\underline{D}_1^k) \text{ and}$$

$$\underline{d}_{1+1}^{k\perp} \in R(\underline{D}_1^k)^\perp \quad . \tag{4.3/14}$$

With this decomposition we obtain from eq. (4.3/13)

$$a_1^k = (\underline{d}_{1+1}^k)^T (\underline{d}_{1+1}^k - \underline{d}_{1+1}^{k''})$$

$$= (\underline{d}_{1+1}^k)^T \underline{d}_{1+1}^{k\perp} = ||\underline{d}_{1+1}^{k\perp}||^2 \quad . \tag{4.3/15}$$

We see that the scalar $a_1^k$ is different from zero if and only if

$$\underline{d}_{1+1}^{k\perp} \neq \underline{0} \tag{4.3/16}$$

is valid, that means if and only if the vector $\underline{d}_{1+1}^k$ is no element of the column space of the matrix $\underline{D}_1^k$. If the condition (4.3/16) is satisfied the vectors $\underline{d}_1^k, \ldots, \underline{d}_1^k$, $\underline{d}_{1+1}^k$ are linearly independent because of the above-mentioned theorem. Therefore it is possible to clear up the question of the linear independence by inspecting the scalar $a_1^k$.

When applying this just derived procedure for examining the linear independence of the vectors $\underline{d}_j^k$ on a digital computer we again have the problem to decide whether a computed value $a_1^k$ is zero or not. But here the advantage of the new method for the solution of nonlinear least squares problems turns out once more. It is not necessary to solve this problem exactly for we can consider the scalar $a_1^k$ from eq. (4.3/15) as zero if it is less than a given positive upper bound $\varepsilon$. The method is stable in any case. But this property can be lost if we make use of the sometimes proposed possibility to determine the direction vector $\underline{r}^k$ with the help of the Moore-Penrose pseudo-inverse (see appendix D) if the column vectors of the matrix $\underline{D}^k$ are nearly linearly dependent. The details concerning this problem can be found in appendix E.

The above-mentioned positive upper bound $\varepsilon$ has to be selected in such a way that we really recognize a linearly dependent vector. We cannot choose the upper bound too small because we must always be aware of numerical errors which cause the scalar $a_1^k$ to be different from zero in the case of linearly dependent vectors. Further conditions for the upper bound $\varepsilon$ are discussed in details in section 4.5.

By the aid of eq. (4.3/15) we can decide whether a further vector is linearly dependent on the already selected vectors or not. In the case that a vector $\underline{d}_{l+1}^k$ is linearly independent on the vectors $\underline{d}_1^k, \ldots, \underline{d}_l^k$ it was pretended that we are able to determine the matrix $((\underline{D}_{l+1}^k)^T \underline{D}_{l+1}^T)^{-1}$ in the main from values which are necessary for the examination of the linear dependence. In order to show this we utilize the relations (4.3/11) in the eqs. (4.3/6). Introducing the vector

$$\underline{u}_l^k = ((\underline{D}_l^k)^T \underline{D}_l^k)^{-1} (\underline{D}_l^k)^T \underline{d}_{l+1}^k = \underline{D}_l^{k+} \underline{d}_{l+1}^k \qquad (4.3/17)$$

we obtain

$$\underline{B}_1 = ((\underline{D}_l^k)^T \underline{D}_l^k)^{-1} + (a_l^k)^{-1} \underline{u}_l^k (\underline{u}_l^k)^T$$

$$\underline{B}_2 = - (a_l^k)^{-1} \underline{u}_l^k$$

$$\underline{B}_3 = - (a_l^k)^{-1} (\underline{u}_l^k)^T$$

$$\underline{B}_4 = (a_l^k)^{-1} \qquad . \qquad (4.3/18)$$

The calculation of the vector $\underline{u}_l^k$ is also necessary for the determination of the scalar $a_l^k$ as we see by inspecting eq. (4.3/13).

With the eqs. (4.3/13), (4.3/17), and (4.3/18) we have an algorithm at our disposal which is able to carry out the successive selection of the vectors $\underline{d}_1^k, \ldots, \underline{d}_l^k$ and which simultaneously yields the matrix $((\underline{D}_l^k)^T \underline{D}_l^k)^{-1}$. The principal components of this selection procedure are shown in the logic diagram 7. At the beginning of the selection procedure it is checked whether a vector $\underline{d}_j^k$ is equal to the null vector respectively whether its absolute value is less than an upper bound $\eta$. In this case it is linearly dependent so that it does not participate in the selection procedure.

Remembering the selection procedure as described above we remark that the selection of the vectors for the building up of the matrix $\underline{D}_l^k$ depends on the random sequence of the vectors $\underline{d}_j^k$. This is illustrated by a simple example. Assume that the two vectors $\underline{d}_1^k$ and $\underline{d}_2^k$ are linearly independent but a further vector $\underline{d}_3^k$ is linearly dependent on the vectors $\underline{d}_1^k$ and $\underline{d}_2^k$. Moreover assume that for the same vectors it is valid that the vectors $\underline{d}_1^k$ and $\underline{d}_3^k$ are linearly independent but that the vector $\underline{d}_2^k$ is linearly dependent on these vectors $\underline{d}_1^k$ and $\underline{d}_3^k$. If we select the vectors according to their random numeration the above described selection procedure yields the vectors $\underline{d}_1^k$ and $\underline{d}_2^k$. But as the numeration of the vectors is arbitrary the selection procedure yields the vectors $\underline{d}_1^k$ and $\underline{d}_3^k$ if we exchange the indices 2 and 3. Therefore it is certainly not very intelligent to select the vectors $\underline{d}_j^k$ for the building up of the matrix $\underline{D}_l^k$ according to the sequence of their examination in the course of the selection procedure. For this reason we shall specify and examine criteria in section 4.4 which are well

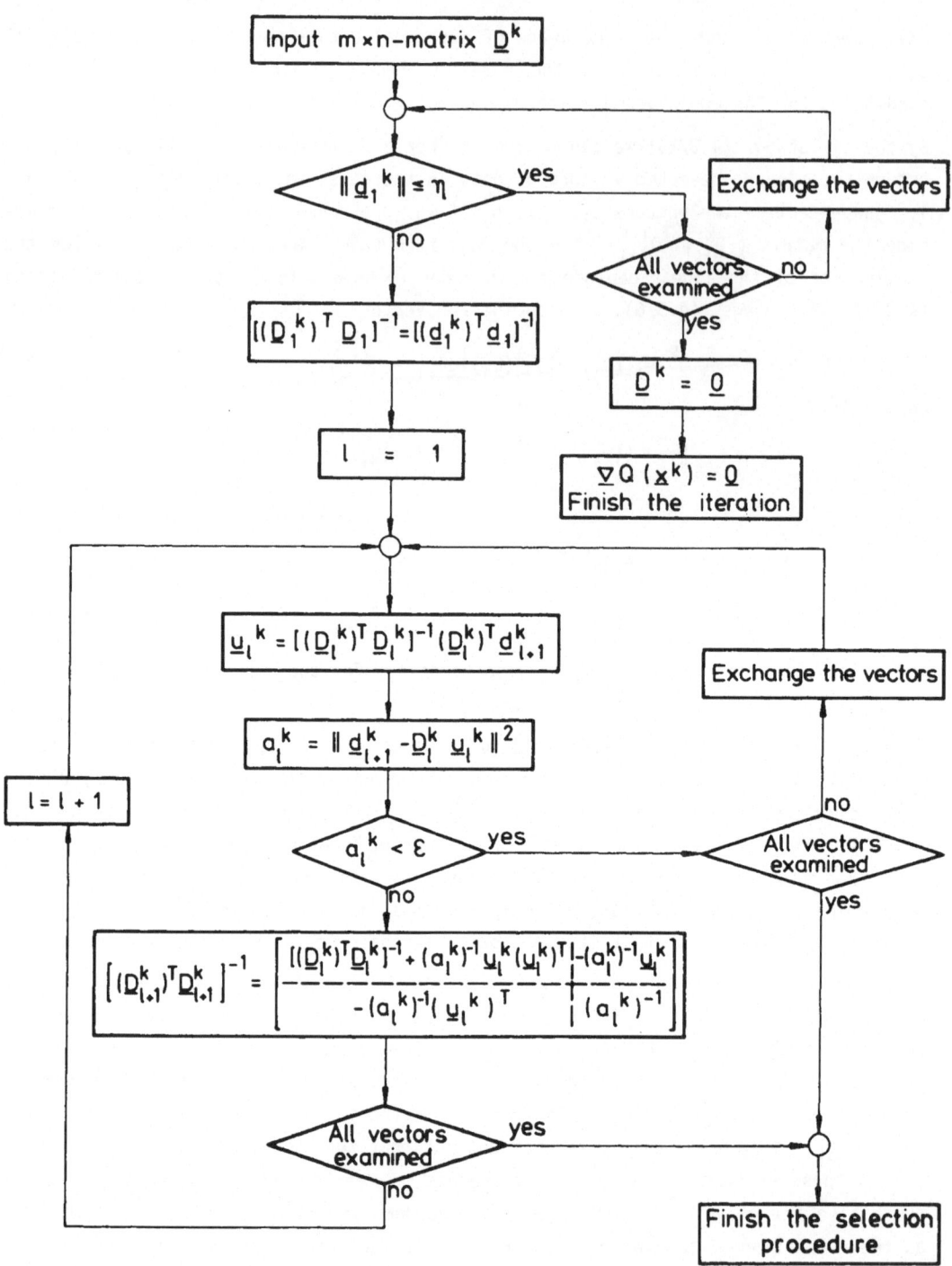

Logic diagram 7: Selection of 1 linearly independent vectors and simultaneous calculation of the matrix $((\underline{D}_1^k)^T\underline{D}_1^k)^{-1}$.

founded and independent on the arbitrary numeration of the vectors $\underline{d}_j^k$.

But before we occupy ourselves with these criteria we want to derive an important relation for orthogonal projectors which is interesting in the sequel [8].

Assume we have selected 1 linearly independent vectors $\underline{d}_1^k$, ..., $\underline{d}_1^k$. Besides it is possible to find a further vector $\underline{d}_{1+1}^k$ so that the vectors $\underline{d}_1^k$, ..., $\underline{d}_1^k$, $\underline{d}_{1+1}^k$ are also linearly independent. Moreover we have a vector $\underline{z} \in \mathbb{R}^m$. On the one hand this vector $\underline{z}$ is decomposed according to

$$\underline{z} = \underline{z}''_{\underline{D}_1^k} + \underline{z}^\perp_{\underline{D}_1^k} \quad \text{with } \underline{z}''_{\underline{D}_1^k} \in R(\underline{D}_1^k), \ \underline{z}^\perp_{\underline{D}_1^k} \in R(\underline{D}_1^k)^\perp. \tag{4.3/19}$$

On the other hand it is decomposed according to

$$\underline{z} = \underline{z}''_{\underline{D}_{1+1}^k} + \underline{z}^\perp_{\underline{D}_{1+1}^k} \quad \text{with } \underline{z}''_{\underline{D}_{1+1}^k} \in R(\underline{D}_{1+1}^k), \ \underline{z}^\perp_{\underline{D}_{1+1}^k} \in R(\underline{D}_{1+1}^k)^\perp. \tag{4.3/20}$$

(The notation $\underline{z}''_{\underline{D}_1^k}$ means that $\underline{z}''_{\underline{D}_1^k}$ is the component of the vector $\underline{z}$ in the column space of the matrix $\underline{D}_1^k$. The meaning of the notation $\underline{z}^\perp_{\underline{D}_1^k}$ is analogous.)

Now we pretend that the inequality

$$\| \underline{z}^\perp_{\underline{D}_{1+1}^k} \| \leq \| \underline{z}^\perp_{\underline{D}_1^k} \| \tag{4.3/21}$$

holds.

To show this we write down the decompositions (4.3/19) and (4.3/20) for the vector $\underline{z}$ by the aid of the orthogonal projectors $\underline{D}_1^k \underline{D}_1^{k+}$ and $\underline{D}_{1+1}^k \underline{D}_{1+1}^{k+}$

$$\underline{z}^\perp_{\underline{D}_1^k} = [\underline{I} - \underline{D}_1^k \underline{D}_1^{k+}] \underline{z} \tag{4.3/22}$$

$$\underline{z}^\perp_{\underline{D}_{1+1}^k} = [\underline{I} - \underline{D}_{1+1}^k \underline{D}_{1+1}^{k+}] \underline{z}$$

$$= [\underline{I} - \underline{D}_{1+1}^k ((\underline{D}_{1+1}^k)^T \underline{D}_{1+1}^k)^{-1} (\underline{D}_{1+1}^k)^T] \underline{z} . \tag{4.3/23}$$

In eq. (4.3/23) we utilize the relation (4.3/18) for the calculation of the matrix $((\underline{D}_{1+1}^k)^T \underline{D}_{1+1}^k)^{-1}$ on the basis of the matrix $((\underline{D}_1^k)^T \underline{D}_1^k)^{-1}$. Then we obtain

$$\underline{z}^\perp_{\underline{D}_{1+1}^k} = [\underline{I} - (\underline{D}_1^k, \underline{d}_{1+1}^k) \begin{pmatrix} ((\underline{D}_1^k)^T \underline{D}_1^k)^{-1} + (a_1^k)^{-1} \underline{u}_1^k (\underline{u}_1^k)^T & - (a_1^k)^{-1} \underline{u}_1^k \\ - (a_1^k)^{-1} (\underline{u}_1^k)^T & (a_1^k)^{-1} \end{pmatrix} \begin{pmatrix} (\underline{D}_1^k)^T \\ (\underline{d}_{1+1}^k)^T \end{pmatrix} ] \underline{z}$$

$$= \{ \underline{I} - \underline{D}_1^k \underline{D}_1^{k+} - (a_1^k)^{-1} [(\underline{I} - \underline{D}_1^k \underline{D}_1^{k+}) \underline{d}_{1+1}^k (\underline{d}_{1+1}^k)^T (\underline{I} - \underline{D}_1^k \underline{D}_1^{k+})] \} \underline{z} \tag{4.3/24}$$

Setting

$$(\underline{I}_1^k)^\perp = \underline{I} - \underline{D}_1^k \underline{D}_1^{k+} \tag{4.3/25}$$

and

$$(\underline{T}_{l+1}^{k})^{\perp} = \underline{I} - \underline{D}_{l+1}^{k}\underline{D}_{l+1}^{k+}$$  (4.3/26)

we obtain from the eqs. (4.3/23) and (4.3/24)

$$(\underline{T}_{l+1}^{k})^{\perp} = (\underline{T}_{l}^{k})^{\perp} - (a_{l}^{k})^{-1}(\underline{T}_{l}^{k})^{\perp}\underline{d}_{l+1}^{k}((\underline{T}_{l}^{k})^{\perp}\underline{d}_{l+1}^{k})^{T}.$$  (4.3/27)

By the aid of this relation we obtain

$$
\begin{aligned}
\|\underline{z}_{\underline{D}_{l+1}^{k}}^{\perp}\|^{2} &= \|(\underline{T}_{l}^{k})^{\perp}\underline{z}\|^{2} + (a_{l}^{k})^{-2}\|((\underline{T}_{l}^{k})^{\perp}\underline{d}_{l+1}^{k})^{T}\underline{z}\|^{2}\|(\underline{T}_{l}^{k})^{\perp}\underline{d}_{l+1}^{k}\|^{2} - \\
&\quad - 2(a_{l}^{k})^{-1}((\underline{T}_{l}^{k})^{\perp}\underline{d}_{l+1}^{k})^{T}\underline{z}((\underline{T}_{l}^{k})^{\perp}\underline{d}_{l+1}^{k})^{T}(\underline{T}_{l}^{k})^{\perp}\underline{z} \\
&= \|\underline{z}_{\underline{D}_{l}^{k}}^{\perp}\|^{2} + (a_{l}^{k})^{-1}\|((\underline{T}_{l}^{k})^{\perp}\underline{d}_{l+1}^{k})^{T}\underline{z}\|^{2} - 2(a_{l}^{k})^{-1}\|((\underline{T}_{l}^{k})^{\perp}\underline{d}_{l+1}^{k})^{T}\underline{z}\|^{2} \\
&= \|\underline{z}_{\underline{D}_{l}^{k}}^{\perp}\|^{2} - (a_{l}^{k})^{-1}\|((\underline{T}_{l}^{k})^{\perp}\underline{d}_{l+1}^{k})^{T}\underline{z}\|^{2} .
\end{aligned}
$$  (4.3/28)

The validity of eq. (4.3/28) is based on eq. (4.3/15) and on the symmetry and idempotency of the matrix $(\underline{T}_{l}^{k})^{\perp}$. Because of $a_{l}^{k} > 0$ we obtain from eq. (4.3/28)

$$\|\underline{z}_{\underline{D}_{l+1}^{k}}^{\perp}\|^{2} \leq \|\underline{z}_{\underline{D}_{l}^{k}}^{\perp}\|^{2}$$  (4.3/29)

which is the pretended inequality.

The inequality (4.3/29) is very important for the selection procedure of the vectors $\underline{d}_{1}^{k}$, ..., $\underline{d}_{l}^{k}$. To see this we assume that we have found a vector $\underline{d}_{j}^{k}$ for which the scalar $a_{l}^{k}$ is less than a given upper bound $\varepsilon$. Until now, it is only clear that this vector $\underline{d}_{j}^{k}$ cannot be selected for the building up of the matrix $\underline{D}_{l}^{k}$ at this moment. But by paying attention to the inequality (4.3/29) we see that this vector can nevermore be selected for the building up of the matrix $\underline{D}_{l}^{k}$ because the scalar $a_{l}^{k}$ referring to the vector $\underline{d}_{j}^{k}$ under consideration will not increase. In other words, this means that a vector $\underline{d}_{j}^{k}$ for which the scalar $a_{l}^{k}$ - calculated in the course of the selection procedure in the stage k of the iteration - is less than an upper bound $\varepsilon$ is eliminated from the further selection procedure of the vectors $\underline{d}_{j}^{k}$ for the building up of the matrix $\underline{D}_{l}^{k}$ in this stage k of the iteration. So it is possible that the number of vectors to be examined is substantially reduced under certain circumstances. This reduction can save computation time.

Based on the inequality (4.3/29) we can make the following considerations concerning the number of vectors $\underline{d}_{j}^{k}$ to be selected for the building up of the matrix $\underline{D}_{l}^{k}$.

As shown in appendix D the vector

$$\underline{r}^{k} = \begin{pmatrix} \underline{D}_{l}^{k+} \\ \underline{0} \end{pmatrix} \underline{e}(\underline{x}^{k})$$  (4.3/30)

is a solution of the system of linear equations

$$\underline{D}^k \underline{r}^k = \underline{e}(\underline{x}^k) \tag{4.3/31}$$

in the sense of

$$||\underline{D}^k \underline{r}^k - \underline{e}(\underline{x}^k)|| \leq ||\underline{D}_1^k \underline{z} - \underline{e}(\underline{x}^k)|| \text{ for all } \underline{z} \in \mathbf{R}^1 \text{ and}$$
$$\underline{D}^k \underline{r}^k \in R(\underline{D}_1^k). \tag{4.3/32}$$

The error when solving the system of linear equations (4.3/31) in the sense of (4.3/32) is given by

$$||\underline{D}^k \underline{r}^k - \underline{e}(\underline{x}^k)|| = ||\underline{e}(\underline{x}^k)^\perp|| \text{ with } \underline{e}(\underline{x}^k)^\perp \in R(\underline{D}_1^k)^\perp. \tag{4.3/33}$$

By inspecting the inequality (4.3/29) we see that this error can decrease under certain circumstances namely if we utilize further linearly independent vectors for the building up of the matrix $\underline{D}_1^k$.

The direction vector $\underline{r}^k$ of the new method for the solution of nonlinear least squares problems can be interpreted as the solution of the system of linear equations (4.3/31) in the sense of the relation (4.3/32) if we substitute the in general non-linear function $\underline{f}(\underline{x})$ by its linear approximation (3.2/1). Based on this linear approximation of the function $\underline{f}(\underline{x})$ we recognize by the inequality (4.3/29) that it can be intelligent to utilize as many linearly independent vectors $\underline{d}_j^k$ as possible for the building up of the matrix $\underline{D}_1^k$ also in the case of nonlinear least squares problems (compare also chapter 5).

In the next section we shall examine further aspects for the selection of the vectors $\underline{d}_j^k$ besides those which have already been considered.

### 4.4 Criteria for the selection of the column vectors for the building up of the matrix $\underline{D}_1^k$

Until now, it is not known which criteria are used for the selection of the vectors $\underline{d}_j^k$ for the building up of the matrix $\underline{D}_1^k$. It is only laid down how to select the first vector namely so that the scalar product (4.3/2) becomes minimal. With this selection the stability of the method is guaranteed when utilizing only one vector for the determination of the direction vector $\underline{r}^k$. Moreover it is known which vectors cannot be used namely all those which are linearly dependent on the already selected vectors. But beyond that the selection procedure is not specified. For the application of the proposed method we are depending on criteria for the selection of the vectors. Two of the possible criteria for the selection will be proposed and discussed in this section. It can be shown that these criteria are well founded.

### 4.4.1 Selection of the vectors for the building up of the matrix $\underline{D}_l^k$ according to the criterion "maximal change of the slope"

Assume we have reached a certain point $\underline{x}^k$ in the parameter space in the course of the search for the point $\underline{x}^*$. A certain value of the sum of squares $Q(\underline{x}^k)$ belongs to this parameter vector $\underline{x}^k$. If we consider the sum of squares $Q(\underline{x})$ along the line $\underline{x} = \underline{x}^k + \alpha \underline{r}^k$ this function $Q(\underline{x}^k + \alpha \underline{r}^k)$ is a function of one variable, namely $\alpha$, whose slope for $\alpha = 0$ - that means at the point $\underline{x}^k$ - is given by

$$\left. \frac{dQ(\underline{x}^k + \alpha \underline{r}^k)}{d\alpha} \right|_{\alpha = 0} = (\nabla Q(\underline{x}^k))^T \underline{r}^k \tag{4.4.1/1}$$

The first vector which is selected for the building up of the matrix $\underline{D}_l^k$ is determined so that the expression in eq. (4.4.1/1) becomes minimal. This is the above described criterion for the selection of the first vector. If we are able to determine the rank of the matrix $\underline{D}^k$ exactly so we can calculate the slope of the function $Q(\underline{x}^k + \alpha \underline{r}^k)$ for $\alpha = 0$ when utilizing all linearly independent vectors for the determination of the direction vector $\underline{r}^k$. Because of eq. (4.2/15) it is given by

$$(\nabla Q(\underline{x}^k))^T \underline{r}^k = -2 \| \underline{e}(\underline{x}^k)'' \|^2 \text{ with } \underline{e}(\underline{x}^k)'' \in R(\underline{D}^k) \ . \tag{4.4.1/2}$$

But as the exact determination of the rank of the matrix $\underline{D}^k$ and with that of the column space of the matrix $\underline{D}^k$ is often not possible on digital computers because of the above-mentioned reasons we cannot calculate the value of the slope from eq. (4.4.1/2) numerically.

Assume we have selected 1 linearly independent vectors $\underline{d}_1^k, \ldots, \underline{d}_l^k$. The slope of the function $Q(\underline{x}^k + \alpha \underline{r}_l^k)$ at the point $\underline{x}^k$ is then given analogously to eq. (4.4.1/2) by

$$(\nabla Q(\underline{x}^k))^T \underline{r}_l^k = -2 \| \underline{e}(\underline{x}^k)''_{\underline{D}_l^k} \|^2 \text{ with } \underline{e}(\underline{x}^k)''_{\underline{D}_l^k} \in R(\underline{D}_l^k) \tag{4.4.1/3}$$

where $\underline{r}_l^k$ is the direction vector determined on the basis of 1 selected vectors $\underline{d}_j^k$.

If we now add a further vector $\underline{d}_{l+1}^k$ which is linearly independent on the already selected vectors $\underline{d}_1^k, \ldots, \underline{d}_l^k$ the slope of the function $Q(\underline{x}^k + \alpha \underline{r}_{l+1}^k)$ at the point $\underline{x}^k$ is given by

$$(\nabla Q(\underline{x}^k))^T \underline{r}_{l+1}^k = -2 \| \underline{e}(\underline{x}^k)''_{\underline{D}_{l+1}^k} \|^2 \text{ with } \underline{e}(\underline{x}^k)''_{\underline{D}_{l+1}^k} \in R(\underline{D}_{l+1}^k). \tag{4.4.1/4}$$

Because of the decomposition

$$\underline{e}(\underline{x}^k) = \underline{e}(\underline{x}^k)'' + \underline{e}(\underline{x}^k)^\perp \tag{4.4.1/5}$$

we have

$$\| \underline{e}(\underline{x}^k) \|^2 = \| \underline{e}(\underline{x}^k)'' \|^2 + \| \underline{e}(\underline{x}^k)^\perp \|^2 \ . \tag{4.4.1/6}$$

Because of the inequality (4.3/29) we obtain

$$\| \underline{e}(\underline{x}^k)^{\perp}_{\underline{D}^k_{l+1}} \|^2 \leq \| \underline{e}(\underline{x}^k)^{\perp}_{\underline{D}^k_l} \|^2 . \qquad (4.4.1/7)$$

Because of eq. (4.4.1/6) it follows from the inequality (4.4.1/7)

$$\| \underline{e}(\underline{x}^k)''_{\underline{D}^k_{l+1}} \|^2 \geq \| \underline{e}(\underline{x}^k)''_{\underline{D}^k_l} \|^2 . \qquad (4.4.1/8)$$

The last inequality means that it is not possible to decrease the orthogonal projection of the error vector $\underline{e}(\underline{x}^k)$ by adding a further linearly independent vector $\underline{d}^k_{l+1}$ to the already selected vectors $\underline{d}^k_1, \ldots, \underline{d}^k_l$. Hence it follows that it is possible to decrease the slope of the function $Q(\underline{x}^k + \alpha \underline{r}^k)$ at the point $\underline{x}^k$ under certain circumstances by adding a further linearly independent vector $\underline{d}^k_{l+1}$.

Therefore we are to select as great a number as possible of linearly independent vectors so that the slope of the function $Q(\underline{x}^k + \alpha \underline{r}^k)$ becomes minimal at the point $\underline{x}^k$. The same requirement results from the remark made at the end of section 4.3 concerning as great a decrease as possible of the sum of squares Q if we substitute the in general nonlinear function $\underline{f}(\underline{x})$ by its linear approximation (3.2/1). The two requirements - concerning as small a slope as possible and as great a decrease as possible of the sum of squares Q - drive at the same direction (for this see also chapter 5).

Depending on these requirements we can develop a strategy for the selection of the next vector.

Among all candidates of vectors which are still under consideration we select that vector which yields as great a decrease as possible of the slope of the function $Q(\underline{x}^k + \alpha \underline{r}^k)$ at the point $\underline{x}^k$. If we perform the selection of the vectors in this way we speak of a selection according to the criterion "maximal change of the slope".

In order to apply the selection procedure on the basis of this criterion without great numerical effort it is necessary to derive a relation between $\| \underline{e}(\underline{x}^k)''_{\underline{D}^k_{l+1}} \|^2$ and $\| \underline{e}(\underline{x}^k)''_{\underline{D}^k_l} \|^2$. This can be done by utilizing the relations (4.3/15), (4.3/17), and (4.3/18). If we write

$$\| \underline{e}(\underline{x}^k)''_{\underline{D}^k_{l+1}} \|^2 = \| \underline{e}(\underline{x}^k)''_{\underline{D}^k_l} \|^2 + \tfrac{1}{4}\Delta^k_l \qquad (4.4.1/9)$$

we obtain after a short calculation

$$\Delta^k_l = (a^k_l)^{-1} [(\underline{\nabla}_l Q(\underline{x}^k))^T \underline{u}^k_l - \frac{\partial Q}{\partial x_{l+1}} \Big|_{\underline{x}_k} ]^2 . \qquad (4.4.1/10)$$

The notation $\underline{\nabla}_l Q(\underline{x}^k)$ in eq. (4.4.1/10) means the first l components of the gradient vector $\underline{\nabla} Q(\underline{x}^k)$. We see that it is possible to calculate the value of $\Delta^k_l$ for each vector $\underline{d}_{l+1}^k$ to be examined practically without any supplementary effort because the scalar $a_l^k$ and the vector $\underline{u}_l^k$ have to be determined for the calculation of the matrix $((\underline{D}_{l+1}^k)^T \underline{D}_{l+1}^k)^{-1}$ anyhow. Considering the logic diagram 6 we see that the vector $\underline{\triangledown}Q(\underline{x}^k)$ is also at our disposal.

Applying the criterion "maximal change of the slope" the selection of the vectors for the building up of the matrix $\underline{D}_l^k$ proceeds in the way that we select that vector next for which the scalar $\Delta_l^k$ from eq. (4.4.1/10) is maximal. Thereby only those vectors are at our disposal for which $a_l^k \neq 0$ respectively $a_l^k > \varepsilon$ is valid. Here it is to remark that it is not necessarily advantageous to examine the question of the linear independence by the aid of the scalar $a_l^k$ but with the help of the angle between the vector $\underline{d}_{l+1}^k$ and the subspace spanned by the vectors $\underline{d}_1^k, \dots, \underline{d}_l^k$ (for more details see subsection 4.4.2). The utilization of the criterion "maximal change of the slope" means that the slope of the function $Q(\underline{x}^k + \alpha\underline{r}^k)$ at the point $\underline{x}^k$ becomes minimal if we only add one further vector to the already selected vectors. By adding further linearly independent vectors we approach to as small a value as possible of the slope of the function $Q(\underline{x}^k + \alpha\underline{r}^k)$ at the point $\underline{x}^k$ which is given by eq. (4.4.1/2).

It can be shown by examples that the sequence of changes of the slope is not monotonously decreasing. This means that when selecting a vector $\underline{d}_l^k$ we can obtain a maximal value for $\Delta_{l-1}^k$ which is less than $\Delta_l^k$ which we obtain when adding a further vector $\underline{d}_{l+1}^k$. Therefore we are not sure that we have sufficiently approached to the theoretically attainable value (4.4.1/2) for the slope of the function $Q(\underline{x}^k + \alpha\underline{r}^k)$ at the point $\underline{x}^k$ if the maximally obtained value $\Delta_l^k$ in one stage of the selection procedure is less than a given positive upper bound.

The application of the described method has shown that it is in general sufficient to stop the selection procedure if the scalar

$$v_l^k = \left| \frac{\max(\Delta_{l-1}^k) - \max(\Delta_l^k)}{\max(\Delta_{l-1}^k)} \right| \qquad (4.4.1/11)$$

is less than a given positive upper bound $\varepsilon_v$ although the sequence of the scalars $v_l^k$ is not monotonously decreasing.

The principal components of the above described selection procedure according to the criterion "maximal change of the slope" are represented in the logic diagram 8. In contrast to the selection procedure represented in the logic diagram 7 the random sequence of the vectors $\underline{d}_j^k$ plays no part. The price for this is that we have to examine each vector $\underline{d}_j^k$ in general several times in the course of the selection procedure with regard to its momentary usefulness for the building up of the matrix $\underline{D}_l^k$. But it cannot happen that a vector $\underline{d}_{l+1}^k$ which yields a great change of the slope is not selected because by chance it is the last vector which is examined and which has to be considered as linearly dependent because the scalar $a_l^k$ is less than a given upper bound $\varepsilon$. This can of course happen because of the relation (4.3/29). For this reason the supplementary effort for this proposed selection procedure is justified.

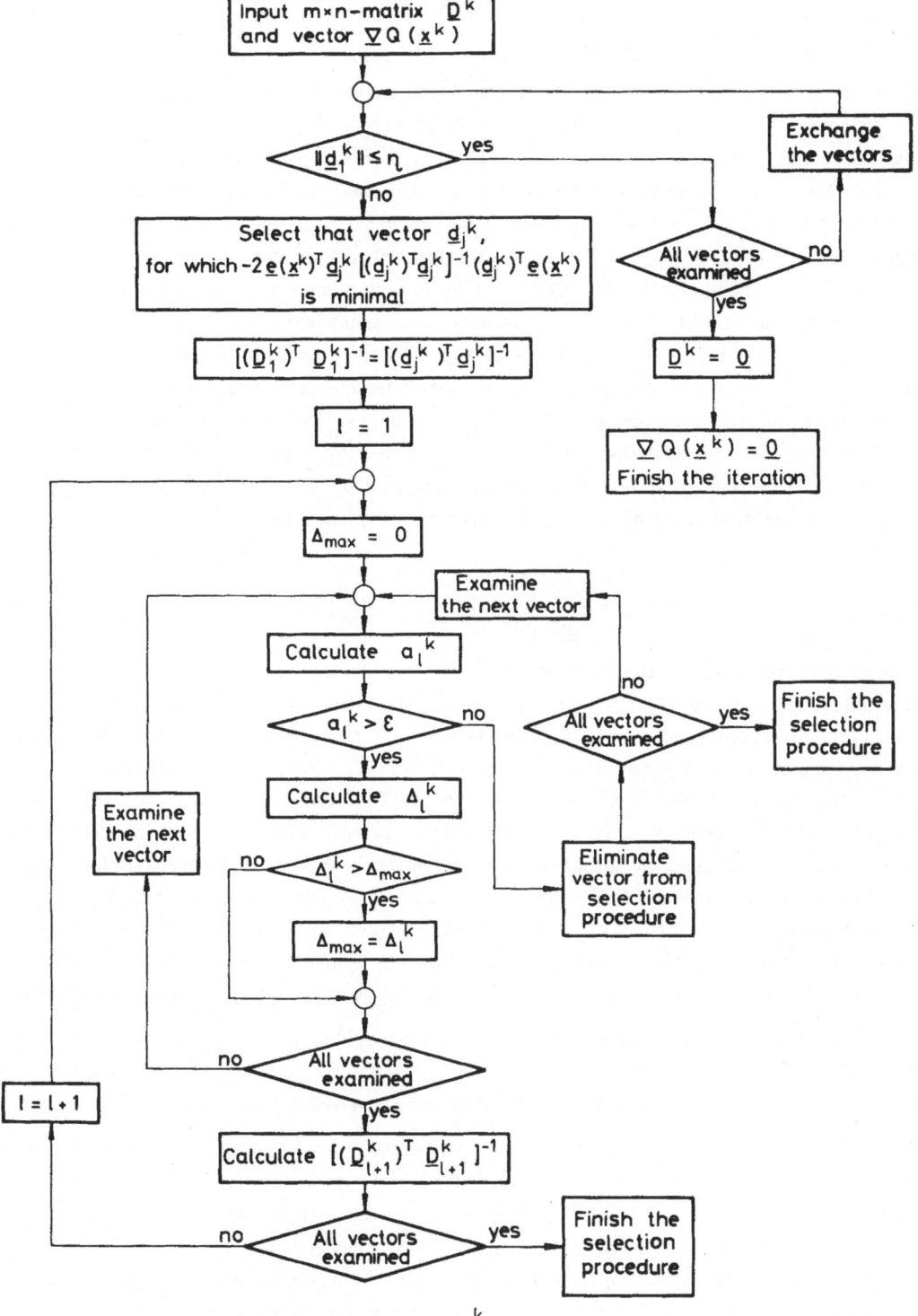

Logic diagram 8: Selection of the vectors $\underline{d}_j^k$ according to the criterion "maximal change of the slope".

### 4.4.2 Selection of the vectors for the building up of the matrix $\underline{D}_1^k$ according to the criterion "maximal angle"

A further possibility for the selection of the column vectors of the matrix consists in selecting that vector $\underline{d}_{1+1}^k$ next which is "most" orthogonal to the subspace spanned by the already selected vectors $\underline{d}_1^k, \ldots, \underline{d}_1^k$. A further vector to be added is the "more" linearly dependent the "less" it is orthogonal to the subspace that means the smaller the absolute value of the angle between this vector and the subspace is. The selection of a vector $\underline{d}_{1+1}^k$ which is nearly linearly dependent on the already selected vectors causes the matrix $(\underline{D}_{1+1}^k)^T \underline{D}_{1+1}^k$ to be nearly singular. The inversion of such a nearly singular matrix is unwelcome because of reasons which are discussed in details in section 4.5. Therefore it is necessary to pay attention to the fact that the vectors to be selected are as little as possible linearly dependent. A measure for the linear independence of a vector $\underline{d}_{1+1}^k$ is not really the scalar $a_1^k$ given by eq. (4.3/15) because the absolute value of the vector $\underline{d}_{1+1}^k$ plays a part but the angle between this vector $\underline{d}_{1+1}^k$ and the subspace spanned by the vectors $\underline{d}_1^k, \ldots, \underline{d}_1^k$. This angle shall be denotated by $\gamma_1$. It is given for instance by

$$\sin\gamma_1 = \frac{\|\underline{d}_{1+1}^{k\perp}\|}{\|\underline{d}_{1+1}^k\|} \text{ with } \underline{d}_{1+1}^{k\perp} \in R(\underline{D}_1^k)^\perp . \tag{4.4.2/1}$$

The selection procedure which makes use of the so defined angle $\gamma_1$ as criterion for the selection proceeds in the way that it selects that vector next for which the scalar $\sin\gamma_1$ is maximal among all vectors which have not yet been selected. This vector is used as next for the building up of the matrix $\underline{D}_1^k$. Then we continue the selection among the not yet selected vectors according to the same criterion. The continuation of this selection procedure is finished if the maximal value of $\sin\gamma_1$ in one stage of the selection procedure is less than a given upper bound $\varepsilon_\gamma$. This examination substitutes the examination of the value of the scalar $a_1^k$ because these two scalars are proportional to each other as can be seen by comparison of the eqs. (4.3/15) and (4.4.2/1). Therefore it is clear that a vector for which the respective angle is less than a given positive upper bound is eliminated from the further selection procedure.

The above described selection procedure shall be called selection of the vectors for the building up of the matrix $\underline{D}_1^k$ according to the criterion "maximal angle". The principal components of this selection procedure are shown in the logic diagram 9.

The scalar $a_1^k$ is a measure for the angle immediately if all vectors $\underline{d}_j^k$ have the same absolute value. In this case it is not necessary to store the absolute values of all vectors $\underline{d}_j^k$, $j = 1, \ldots n$ and to divide the vectors by their absolute values in each stage of the selection procedure. Therefore it is convenient to scale the vectors $\underline{d}_j^k$ before we begin the selection of the vectors for the building up of the matrix $\underline{D}_1^k$. It is suitable to choose the absolute value of the scaled vectors as 1. This scaling is often utilized when solving linear least squares problems that means such problems

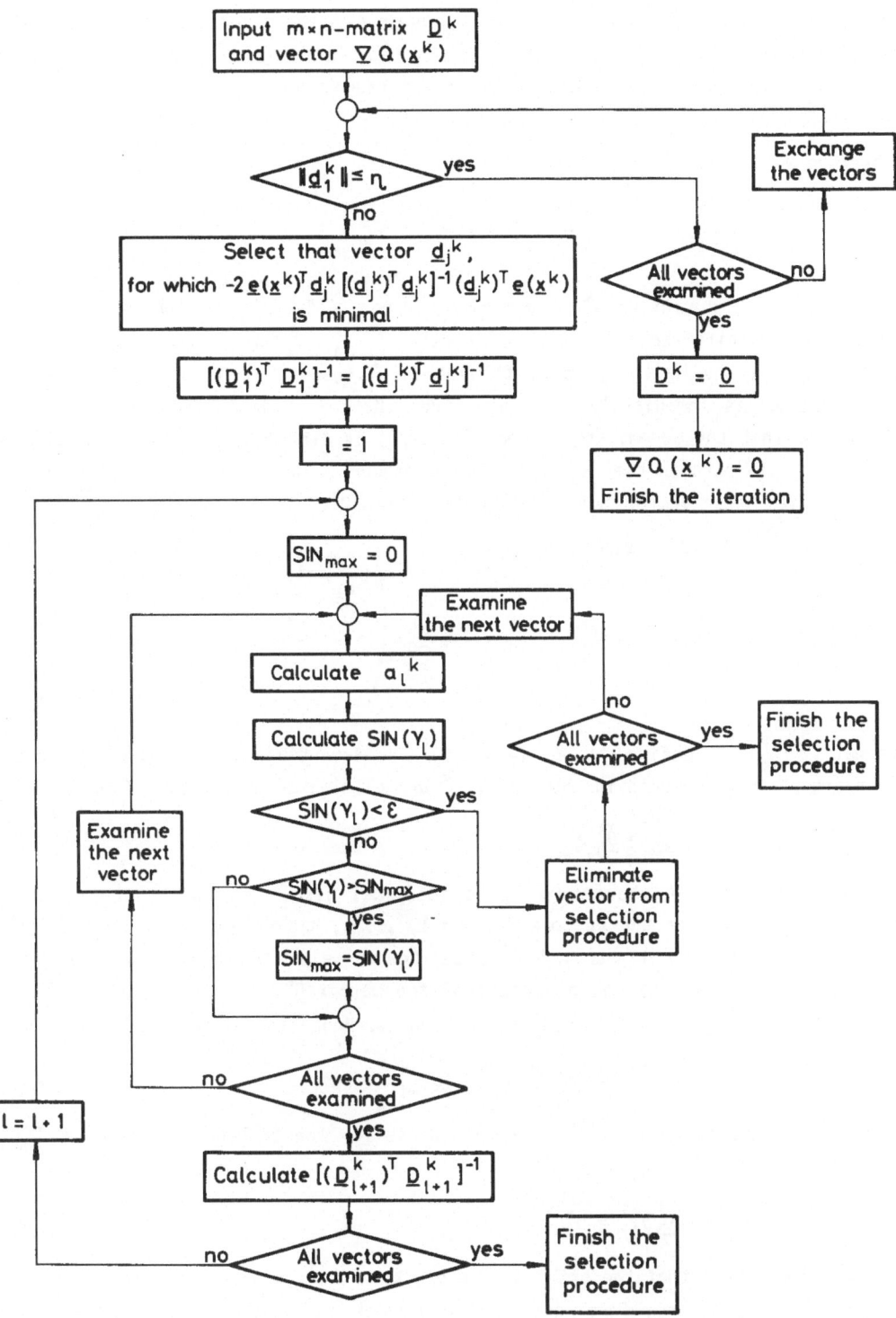

Logic diagram 9: Selection of the vectors $\underline{d}_j^k$ according to the criterion "maximal angle".

for which the function $\underline{f}(\underline{x})$ only depends linearly on the parameter vector $\underline{x}$ which is to be determined [2, 9]. In [10] it is mentioned that the scaling can significantly improve the accuracy of the calculations and thereby can cause a reduction of the number of iterations also in the case of nonlinear least squares problems.

Therefore we introduce the scaled vectors

$$\underline{\tilde{d}}_j^k = \frac{\underline{d}_j^k}{||\underline{d}_j^k||} \, , \, j = 1, \, \ldots, \, n \tag{4.4.2/2}$$

besides the column vectors $\underline{d}_j^k$. The scaling in eq. (4.4.2/2) is only possible if all vectors $\underline{d}_j^k$ are different from the null vector. If $\underline{d}_j^k = \underline{0}$ holds, such a vector is not appropriate to the building up of the matrix $\underline{D}_l^k$ because it is linearly dependent. Such a vector is immediately eliminated from the selection procedure. Therefore we can assume that the denominator in eq. (4.4.2/2) is different from zero for all vectors which take part in the real selection procedure. These are $q \le n$ vectors $\underline{d}_1^k$, $\ldots$, $\underline{d}_q^k$. With these vectors we build up the matrix $\underline{D}_q^k$.

Because of eq. (4.4.2/2) we obtain

$$\underline{\tilde{D}}_q^k = (\underline{\tilde{d}}_1^k, \, \ldots, \, \underline{\tilde{d}}_q^k) = (\underline{d}_1^k, \, \ldots, \, \underline{d}_q^k) \begin{pmatrix} ||\underline{d}_1^k||^{-1} & & \\ & \ddots & \\ & & ||\underline{d}_q^k||^{-1} \end{pmatrix}$$

$$= \underline{D}_q^k (\underline{\Delta}_q^k)^{-1} \tag{4.4.2/3}$$

The matrix $(\underline{\Delta}_q^k)^{-1}$ - defined by eq. (4.4.2/3) - is invertible because the null vectors which are possibly on hand in the matrix $\underline{D}^k$ are eliminated previously. Therefore

$$\underline{D}_q^k = \underline{\tilde{D}}_q^k \, \underline{\Delta}_q^k \tag{4.4.2/4}$$

is valid. The rank of the matrix $\underline{\tilde{D}}_q^k$ is identical with the rank of the matrix $\underline{D}_q^k$ because the scaling of the vectors $\underline{d}_j^k$, $j = 1, \, \ldots \, q$, has no influence on the linear dependence respectively linear independence. Now we select a subset of 1 linearly independent vectors out of the q vectors of the matrix $\underline{D}_q^k$. Then the corresponding 1 vectors $\underline{\tilde{d}}_1^k, \, \ldots, \, \underline{\tilde{d}}_l^k$ are also linearly independent. Hence it follows

$$\underline{D}_l^{k+} = [(\underline{\tilde{D}}_l^k \underline{\Delta}_l^k)^T \underline{\tilde{D}}_l^k \underline{\Delta}_l^k]^{-1} (\underline{\tilde{D}}_l^k \underline{\Delta}_l^k)^T = (\underline{\Delta}_l^k)^{-1} \underline{\tilde{D}}_l^{k+} \tag{4.4.2/5}$$

By paying attention to eq. (4.4.2/4) we obtain for the orthogonal projector $\underline{D}_l^k \underline{D}_l^{k+}$ from eq. (4.4.2/5)

$$\underline{D}_l^k \underline{D}_l^{k+} = \underline{\tilde{D}}_l^k \underline{\tilde{D}}_l^{k+} \, . \tag{4.4.2/6}$$

Eq. (4.4.2/6) means that the orthogonal projector $\underline{D}_l^k \underline{D}_l^{k+}$ is the same referring to the vectors $\underline{d}_1^k, \, \ldots, \, \underline{d}_l^k$ as referring to the vectors $\underline{\tilde{d}}_1^k, \, \ldots, \, \underline{\tilde{d}}_l^k$. Hence it follows that the angle between a vector $\underline{\tilde{d}}_{l+1}^k$ and the subspace spanned by the vectors $\underline{\tilde{d}}_1^k, \, \ldots, \, \underline{\tilde{d}}_l^k$ is identical with the angle between the corresponding vector $\underline{d}_{l+1}^k$ and the subspace spanned by the vectors $\underline{d}_1^k$, ..., $\underline{d}_k^k$. This means that the selection of the vectors in the system $\sim$ yields the same sequence of selected vectors as the selection in the original system. So it is clear that it is arbitrary whether we perform the selection of the vectors for the building up of the matrix $\underline{D}_1^k$ in the system $\sim$ or in the original system.

This is also true for the first vector to be selected. This vector is determined in such a way that the right-hand side of eq. (4.3/2) becomes minimal. We see immediately that the expression on the right-hand side of eq. (4.3/2) is invariant under the transformation (4.4.2/2).

The same is true for the selection of the vectors according to the criterion "maximal change of the slope". Utilizing this criterion the component of the error vector $\underline{e}(\underline{x}^k)$ in the column space of the matrix $\underline{D}_1^k$ is decisive. But this component is invariant under the scaling of the vectors $\underline{d}_j^k$ because of eq. (4.4.2/6).

For the above-mentioned reasons it is intelligent to perform the selection of the vectors also in the system $\sim$ if we utilize the criterion "maximal change of the slope" for the selection because we directly have a measure for the angle $\gamma_1$ when calculating the scalar $a_1^k$.

It is convenient to calculate the matrix $((\underline{D}_1^k)^T \underline{D}_1^k)^{-1}$ also in the system $\sim$. By utilizing eq. (4.4.2/4) we obtain

$$((\underline{D}_1^k)^T \underline{D}_1^k)^{-1} = [(\underline{\tilde{D}}_1^k \underline{\Delta}_1^k)^T \underline{\tilde{D}}_1^k \underline{\Delta}_1^k]^{-1} = (\underline{\Delta}_1^k)^{-1}((\underline{\tilde{D}}_1^k)^T \underline{\tilde{D}}_1^k)^{-1}(\underline{\Delta}_1^k)^{-1}. \qquad (4.4.2/7)$$

We recognize that we obtain the matrix $((\underline{D}_1^k)^T \underline{D}_1^k)^{-1}$ we are looking for if we multiply the matrix $((\underline{\tilde{D}}_1^k)^T \underline{\tilde{D}}_1^k)^{-1}$ by $(\underline{\Delta}_1^k)^{-1}$ from the right- and left-hand side. But we need not perform these two multiplications if we remember that we are not essentially interested in the matrix $((\underline{D}_1^k)^T \underline{D}_1^k)^{-1}$ but only in that part $\underline{r}_1^k$ of the direction vector $\underline{r}^k$ which can be different from the null vector. Because of eq. (4.2/34) this vector $\underline{r}_1^k$ is given by

$$\underline{r}_1^k = (\underline{\Delta}_1^k)^{-1}((\underline{\tilde{D}}_1^k)^T \underline{\tilde{D}}_1^k)^{-1}(\underline{\Delta}_1^k)^{-1}\underline{\Delta}_1^k(\underline{\tilde{D}}_1^k)^T \underline{e}(\underline{x}^k)$$

$$= (\underline{\Delta}_1^k)^{-1}((\underline{\tilde{D}}_1^k)^T \underline{\tilde{D}}_1^k)^{-1}(\underline{\tilde{D}}_1^k)^T \underline{e}(\underline{x}^k) = (\underline{\Delta}_1^k)^{-1}\underline{\tilde{r}}_1^k . \qquad (4.4.2/8)$$

By inspecting eq. (4.4.2/8) we see that it is only necessary to multiply the direction vector $\underline{\tilde{r}}_1^k$ calculated in the system $\sim$ by $(\underline{\Delta}_1^k)^{-1}$ from the left-hand side. So we obtain the direction vector $\underline{r}_1^k$ in the original system. As the matrix $(\underline{\Delta}_1^k)^{-1}$ is a diagonal matrix the numerical effort is minimal to calculate the direction vector $\underline{r}_1^k$ on the basis of the direction vector $\underline{\tilde{r}}_1^k$.

Now we summarize shortly how the proceeding is if we want to calculate the part $\underline{r}_1^k$ of the direction vector $\underline{r}^k$ which can be different from the null vector if we utilize the scaling (4.4.2/2) for the vectors $\underline{d}_j^k$ and if we select the vectors for the building up of the matrix $\underline{D}_1^k$ according to the criterion "maximal change of slope" or according to the criterion "maximal angle". To begin with the vectors $\underline{d}_j^k$ which are equal to the null vector or the absolute values of which are less than a given positive upper bound $\eta$ are eliminated. Assume these are the vectors $\underline{d}_{q+1}^k, \ldots, \underline{d}_n^k$. The remaining vectors $\underline{d}_1^k, \ldots, \underline{d}_q^k$ are gathered up in the matrix $\underline{D}_q^k$ which is transformed according to eq. (4.4.2/3). The selection of the vectors takes place in the system $\sim$ according to one of the two above described criteria for the selection. The calculation of the part of the direction vector $\underline{r}^k$ which can be different from the null vector is also performed in the system $\sim$. At the end the inverse transformation (4.4.2/8) is carried out in order to obtain the direction vector $\underline{r}^k$. The principal components of this procedure are shown in the logic diagram 10.

### 4.4.3  Comparison of the criteria "maximal change of the slope" and "maximal angle"

In the two preceding subsections we have presented two criteria for the selection of the vectors $\underline{d}_j^k$ for the building up of the matrix $\underline{D}_1^k$. An obvious question is which of the two criteria for the selection is more efficient in connexion with the proposed new method for the solution of nonlinear least squares problems. To begin with we have to decide when we want to consider the resulting method as more efficient. Usual criteria for such a comparison are the necessary storage and the computation time which is needed to solve a nonlinear least squares test problem. As to the storage the proposed method with the criterion "maximal change of the slope" does not differ from the method with the criterion "maximal angle". From this point of view none of the two criteria has an advantage.

As to the comparison of the computation times we can state that it is nearly equal for the two criteria in each stage of the iteration if we assume that the same number of vectors is selected in each stage of the iteration which is often true. The equality of the computation times results from the nearly equal number of arithmetic operations and storage accesses. Therefore it is sufficient for a rough comparison to use the number of iteration steps - necessary to find the parameter vector $\underline{x}^*$ - as a measure for the efficiency of the new method for the solution of nonlinear least squares problems when utilizing the one or other criterion for the selection of the vectors $\underline{d}_j^k$.

A comparison which is generally valid cannot be given because the performance of the method can only be investigated on digital computers for a finite number of functions $\underline{f}(\underline{x})$ and for each function only for a finite number of starting points $\underline{x}^0$. Such investigations have been performed for a lot of examples. The examples under consideration arise from problems in the field of control theory. Some of these examples together with the functions $\underline{f}(\underline{x})$ are discussed in details in chapter 7. The result of these pretty extensive investigations can be summarized as follows. It turns out that the proposed new method in connexion with the criterion "maximal change of the slope" in general needs a lower number of iteration steps than the proposed method in

63

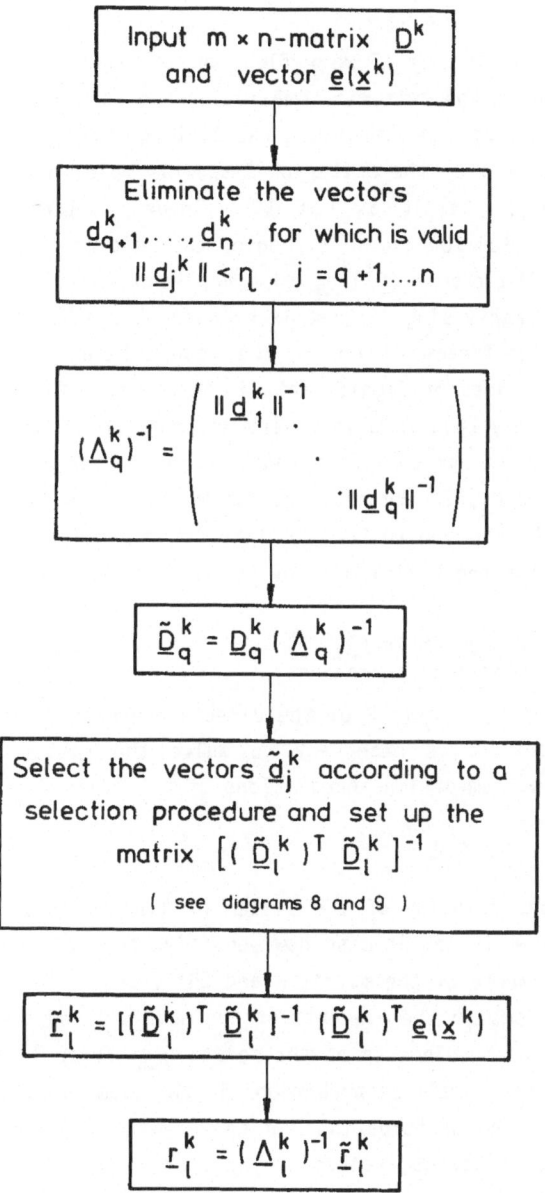

Logic diagram 10: Determination of the vector $\underline{r}_l^k$ by utilizing the transformation (4.4.2/3).

connexion with the criterion "maximal angle". When applying the latter criterion it happened sometimes that the point $\underline{x}^*$ was not reached during the given computation time although the point $\underline{x}^*$ was reached during this computation time when applying the criterion "maximal change of the slope". Hence it follows that it is convenient to select the vectors for the building up of the matrix $\underline{D}_1^k$ according to the criterion "maximal change of the slope" when paying attention to as short a computation time as possible.

This result is not surprising if we reflect on the fact that the selection of the vectors $\underline{d}_j^k$ according to the criterion "maximal change of the slope" aims at as great a decrease as possible of the sum of squares Q. In contrast to this the selection of the vectors $\underline{d}_j^k$ according to the criterion "maximal angle" aims at the property that the vectors $\underline{d}_j^k$ are as little as possible linearly dependent. By utilizing the criterion "maximal angle" for the selection of the vectors $\underline{d}_j^k$ we pay attention to the point of view that the matrix $(\underline{D}_1^k)^T \underline{D}_1^k$ does not become nearly singular. It shall be avoided to invert a nearly singular matrix because it can happen for instance that we obtain a loss of significant digits in the result because of the substraction of great numbers. Such a loss of significant digits can falsify the result of a numerical inversion completely [4]. This is a very important point of view. We must pay attention to it also if we use the criterion "maximal change of the slope" for the selection of the vectors $\underline{d}_j^k$. In the next section we shall discuss this point of view in details and we shall show how to take it into consideration simultaneously with the selection procedure for the building up of the matrix $\underline{D}_1^k$.

## 4.5   The condition of the matrix $(\underline{D}_1^k)^T \underline{D}_1^k$

In order to determine that part $\underline{r}_1^k$ of the direction vector $\underline{r}^k$ which can be different from the null vector it is necessary to solve the system of linear equations (4.2/35). This is a system of linear equations of the type

$$\underline{A}\,\underline{z} = \underline{b}\;. \tag{4.5/1}$$

In connexion with the solution of the system of linear equations (4.5/1) it is of great interest to know if and in case how sensitive the solution $\underline{z}$ is to small perturbations in the elements of the matrix $\underline{A}$ and the vector $\underline{b}$ [11, 12]. It is desirable that small perturbations in the elements of the matrix $\underline{A}$ and the vector $\underline{b}$ only cause small perturbations in the elements of the solution $\underline{z}$. Several examples in [3] show that it is possible that small perturbations in the elements of the matrix $\underline{A}$ and the vector $\underline{b}$ can cause large perturbations in the elements of the solution $\underline{z}$. This is always possible if the absolute values of the elements of the matrix $\underline{A}^{-1}$ are large. This is the case, if the absolute value of the determinant of the matrix $\underline{A}$ is small. But the absolute value of the determinant of the matrix $\underline{A}$ is not solely decisive for the behaviour of the solution $\underline{z}$ if we have small perturbations in the elements of the matrix $\underline{A}$ and the vector $\underline{b}$ [1, 3].

If we have small perturbations in the elements of the matrix $\underline{A}$ and the vector $\underline{b}$ the magnitude of the perturbations in the elements of the solution $\underline{z}$ depends on the matrix $\underline{A}$ of the system of linear equations (4.5/1). Therefore it is possible to divide matrices into those which have the property that small perturbations in the elements of the matrix $\underline{A}$ and the vector $\underline{b}$ cause small perturbations in the solution $\underline{z}$ and into those which do not possess this property. Matrices with this property are called well-conditioned, the others are called ill-conditioned [12]. In this connexion we also speak of the condition of the matrix $\underline{A}$. Naturally the division into well-conditioned and ill-conditioned matrices is very rough and not uniquely determined.

If it is necessary to solve a system of linear equations of the type (4.5/1) in the course of a numerical method we are interested that the perturbations in the elements of the solution $\underline{z}$ are only small if the perturbations in the elements of the matrix $\underline{A}$ and the vector $\underline{b}$ are small that means we are interested in a well-conditioned matrix $\underline{A}$. But mostly we have no influence on the matrix $\underline{A}$ to be inverted, because it is given. If we know that it is ill-conditioned we only have the possibility to utilize a well appropriate numerical method for its inversion that means a method which pays regard to this property of the matrix $\underline{A}$. In spite of this measure we have to consider the solution of the system of linear equations with caution because the original property of a matrix $\underline{A}$ cannot be modified by a numerical method.

If we solve a system of linear equations (4.5/1) on a digital computer the perturbations in the elements of the matrix $\underline{A}$ and the vector $\underline{b}$ result from the finite number of bits for the representation of a real number and essentially from the inevitable rounding errors. In order to keep the influence of these errors on the solution $\underline{z}$ small we prefer a well-conditioned matrix $\underline{A}$. But we do not know a priori if a matrix $\underline{A}$ has this property. In order to examine this we are interested in a number which is a quantitative measure for it. There is a lot of such numbers which are called condition numbers [1, 3, 4]. These condition numbers can be determined with more or less numerical effort. Often the condition number

$$\varkappa(\underline{A}) = \sqrt{\frac{\mu_1}{\mu_n}} \tag{4.5/2}$$

is used. In eq. (4.5/2) $\mu_1$ is the largest and $\mu_n$ is the smallest eigenvalue of the matrix $\underline{A}^T\underline{A}$ [13]. This number $\varkappa(\underline{A})$ is also defined for a nonquadratic matrix $\underline{A}$. By inspecting eq. (4.5/2) we see that the condition number $\varkappa(\underline{A})$ ranges from 1 to $\infty$, because the matrix $\underline{A}^T\underline{A}$ only possesses real non-negative eigenvalues [4]. An ill-conditioned matrix $\underline{A}$ is characterized by a large condition number $\varkappa(\underline{A})$.

Now we want to determine the condition number $\varkappa$ for the matrix which has to be inverted in the course of the proposed new method for the solution of nonlinear least squares problems. For that purpose we assume that we have selected 1 linearly independent vectors $\underline{d}_1^k$, ..., $\underline{d}_1^k$ for the building up of the matrix $\underline{D}_1^k$. The matrix which has to be inverted is $(\underline{D}_1^k)^T\underline{D}_1^k$. For the condition number of this matrix the relation

$$\varkappa((\underline{D}_l^k)^T\underline{D}_l^k) = [\varkappa(\underline{D}_l^k)]^2 \qquad (4.5/3)$$

is valid [13]. By inspecting eq. (4.5/3) we see that the condition number of the matrix $(\underline{D}_l^k)^T\underline{D}_l^k$ is the square of the condition number of the matrix $\underline{D}_l^k$. That means that a numerical method which utilizes the matrix $(\underline{D}_l^k)^T\underline{D}_l^k$ in order to determine the direction vector $\underline{r}^k$ is ill appropriate if the matrix $\underline{D}_l^k$ is ill-conditioned.

Assume, $\mu_l$ is the smallest and $\mu_l$ is the largest eigenvalue of the matrix $(\underline{D}_l^k)^T\underline{D}_l^k$. Hence it follows from eq. (4.5/3)

$$\varkappa((\underline{D}_l^k)^T\underline{D}_l^k) = \frac{\mu_l}{\mu_l} \quad . \qquad (4.5/4)$$

We see that it is necessary to determine the largest and smallest eigenvalue of the matrix $(\underline{D}_l^k)^T\underline{D}_l^k$ in order to examine the condition of this matrix. This determination of the eigenvalues requires a pretty considerable numerical effort. Moreover it is not easy to determine the eigenvalues accurately to a certain extent if the matrix $(\underline{D}_l^k)^T\underline{D}_l^k$ is ill-conditioned. For these reasons we shall not pursue the calculation of the condition number $\varkappa$ in the sequel. But it is possible to give an estimation of the condition number $\varkappa$ of the matrix $(\underline{D}_{l+1}^k)^T\underline{D}_{l+1}^k$ if we know the condition number $\varkappa$ belonging to the matrix $(\underline{D}_l^k)^T\underline{D}_l^k$. Assume that the matrix $\underline{D}^k$ has the theoretical rank $r$. Therefore it can be decomposed according to

$$\underline{D}^k = (\underline{D}_r^k, \ \underline{D}_{n-r}^k) \quad . \qquad (4.5/5)$$

The matrix $\underline{D}_r^k$ is built up by a subset of $r$ linearly independent column vectors. Hence the matrix $\underline{D}_{n-r}^k$ can be obtained by a linear transformation from the matrix $\underline{D}_r^k$. We select $l$ linearly independent column vectors out of the $r$ linearly independent column vectors of the matrix $\underline{D}_r^k$. With these $l$ vectors we build up the matrix $\underline{D}_l^k$. Then the inequality

$$\varkappa(\underline{D}_l^k) \leq \varkappa(\underline{D}_r^k) \qquad (4.5/6)$$

holds [13]. Referring to the selection procedure of the vectors for the building up of the matrix $\underline{D}_l^k$ this inequality means that the condition number of the matrix to be inverted does not decrease but in general increases if we add further linearly independent vectors for the building up of the matrix $\underline{D}_l^k$. In order to avoid an ill-conditioned matrix $\underline{D}_l^k$ it is not always intelligent to use all linearly independent vectors for the building up of the matrix $\underline{D}_l^k$. But as we know this is not necessary if we use the proposed new method for the solution of nonlinear least squares problems.

Because of the pretty considerable numerical effort and of the difficulty to determine the eigenvalues of an ill-conditioned matrix we want to use another condition number instead of the number $\varkappa$ given by eq. (4.5/4) in order to examine the condition of the matrix to be inverted. This condition number which will be used in the sequel can be computed more easily. For a nxn-matrix $\underline{A}$ it is defined by

$$\varkappa_H = \frac{V}{|\det \underline{A}|} \tag{4.5/7}$$

[4] whereby V is given by

$$V = \prod_{j=1}^{n} \|\underline{a}_{j}^{T}\| \ . \tag{4.5/8}$$

In eq. (4.5/8) we denotate the j-th row vector of the matrix $\underline{A}$ by $\underline{a}_j^T$. The condition number $\varkappa_H$ given by eq. (4.5/7) has the following properties [4]:

1. $\varkappa_H$ ranges from 1 to $\infty$.

2. $\varkappa_H$ is invariant under permutations of the row vectors of the matrix $\underline{A}$.

3. $\varkappa_H$ is invariant under multiplications of the row vectors of the matrix $\underline{A}$ with scalars $p_j \neq 0$.

In the sequel we want to judge the condition of the matrix $(\underline{D}_l^k)^T\underline{D}_l^k$ by the aid of the condition number $\varkappa_H$. Because of the above-mentioned reasons it is convenient to use the transformed matrix $\widetilde{\underline{D}}_l^k$ from eq. (4.4.2/3) instead of the matrix $\underline{D}_l^k$. Soon it will be clear that the determination of the condition number $\varkappa_H$ is comparatively simple when performing it in the system $\sim$.

As can be seen by eq. (4.5/7) it is necessary to calculate the value V and $|\det((\widetilde{\underline{D}}_l^k)^T\widetilde{\underline{D}}_l^k)|$ . As the matrix $\widetilde{\underline{D}}_l^k$ is built up by a successive adding of further linearly independent vectors it is suitable to calculate the values necessary for the evaluation of eq. (4.5/7) also successively. For that purpose we show how we can calculate the value of the determinant of the matrix $(\widetilde{\underline{D}}_{l+1}^k)^T\widetilde{\underline{D}}_{l+1}^k$ on the basis of the value of the determinant of the matrix $(\widetilde{\underline{D}}_l^k)^T\widetilde{\underline{D}}_l^k$.

Assume we have selected l linearly independent vectors $\widetilde{\underline{d}}_1^k$, ..., $\widetilde{\underline{d}}_l^k$ which are gathered up in the matrix $\widetilde{\underline{D}}_l^k$. Then the matrix $(\widetilde{\underline{D}}_l^k)^T\widetilde{\underline{D}}_l^k$ is given by

$$(\widetilde{\underline{D}}_l^k)^T\widetilde{\underline{D}}_l^k = \begin{pmatrix} (\widetilde{\underline{d}}_1^k)^T\widetilde{\underline{d}}_1^k & \cdots & (\widetilde{\underline{d}}_1^k)^T\widetilde{\underline{d}}_l^k \\ \vdots & & \vdots \\ (\widetilde{\underline{d}}_l^k)^T\widetilde{\underline{d}}_1^k & \cdots & (\widetilde{\underline{d}}_l^k)^T\widetilde{\underline{d}}_l^k \end{pmatrix} \ . \tag{4.5/9}$$

The matrix defined by eq. (4.5/9) is called the Gram matrix [14] and is denotated by

$$\underline{G}(\widetilde{\underline{d}}_1^k, \ldots, \widetilde{\underline{d}}_l^k) = (\widetilde{\underline{D}}_l^k)^T\widetilde{\underline{D}}_l^k \ . \tag{4.5/10}$$

The determinant belonging to the Gram matrix $\underline{G}(\widetilde{\underline{d}}_1^k, \ldots, \widetilde{\underline{d}}_l^k)$ is the so-called Gram determinant which is denotated by

$$g(\widetilde{\underline{d}}_1^k, \ldots, \widetilde{\underline{d}}_l^k) = \det[\underline{G}(\widetilde{\underline{d}}_1^k, \ldots, \widetilde{\underline{d}}_l^k)] \ . \tag{4.5/11}$$

For the following considerations we need the theorem given in [14]:

Given an arbitrary point $\underline{z} \in \mathbf{R}^m$ and l linearly independent vectors $\widetilde{\underline{d}}_1^k$, ..., $\widetilde{\underline{d}}_l^k \in IR^m$.

The minimum distance $\delta$ of the point $\underline{z}$ to the subspace $R(\underline{\widetilde{D}}_1^k)$ spanned by the vectors $\underline{\widetilde{d}}_1^k, \ldots, \underline{\widetilde{d}}_1^k$ is given by

$$\delta^2 = \frac{g(\underline{\widetilde{d}}_1^k, \ldots, \underline{\widetilde{d}}_1^k, \underline{z})}{g(\underline{\widetilde{d}}_1^k, \ldots, \underline{\widetilde{d}}_1^k)} \tag{4.5/12}$$

If we decompose the vector $\underline{z}$ according to

$$\underline{z} = \underline{z}'' + \underline{z}^{\perp} \text{ with } \underline{z}'' \in R(\underline{\widetilde{D}}_1^k) \text{ and } \underline{z}^{\perp} \in R(\underline{\widetilde{D}}_1^k)^{\perp} \tag{4.5/13}$$

so we have

$$\|\underline{z}^{\perp}\| = \delta^2 . \tag{4.5/14}$$

In the special case that we set the vector $\underline{z}$ equal to the vector $\underline{\widetilde{d}}_{1+1}^k$ it follows from eq. (4.5/12) in connexion with eq. (4.5/14)

$$g(\underline{\widetilde{d}}_1^k, \ldots, \underline{\widetilde{d}}_1^k, \underline{\widetilde{d}}_{1+1}^k) = \|\underline{\widetilde{d}}_{1+1}^{k\perp}\|^2 g(\underline{\widetilde{d}}_1^k, \ldots, \underline{\widetilde{d}}_1^k) . \tag{4.5/15}$$

Utilizing the relation (4.3/15) we can write for eq. (4.5/15)

$$g(\underline{\widetilde{d}}_1^k, \ldots, \underline{\widetilde{d}}_1^k, \underline{\widetilde{d}}_{1+1}^k) = \widetilde{a}_1^k g(\underline{\widetilde{d}}_1^k, \ldots, \underline{\widetilde{d}}_1^k) . \tag{4.5/16}$$

Instead of the scalar $a_1^k$ we have to use the scalar $\widetilde{a}_1^k$ because we consider the selection of the vectors $\underline{d}_j^k$ in the system $\sim$. Applying the relation (4.5/15) to $g(\underline{\widetilde{d}}_1^k, \ldots, \underline{\widetilde{d}}_1^k)$ in an analogous way we obtain from eq. (4.5/16)

$$g(\underline{\widetilde{d}}_1^k, \ldots, \underline{\widetilde{d}}_1^k, \underline{\widetilde{d}}_{1+1}^k) = \widetilde{a}_1^k \widetilde{a}_{1-1}^k g(\underline{\widetilde{d}}_1^k, \ldots, \underline{\widetilde{d}}_{1-1}^k) . \tag{4.5/17}$$

By repeated application of the relation (4.5/15) we finally obtain

$$g(\underline{\widetilde{d}}_1^k, \ldots, \underline{\widetilde{d}}_1^k, \underline{\widetilde{d}}_{1+1}^k) = \prod_{j=1}^{1} \widetilde{a}_j^k g(\underline{\widetilde{d}}_1^k) . \tag{4.5/18}$$

As the vector $\underline{\widetilde{d}}_1^k$ is scaled in such a way that its Euclidean length is 1 its Gram determinant is given by

$$g(\underline{\widetilde{d}}_1^k) = 1 . \tag{4.5/19}$$

Paying attention to this equality we obtain from eq. (4.5/18)

$$g(\underline{\widetilde{d}}_1^k, \ldots, \underline{\widetilde{d}}_1^k, \underline{\widetilde{d}}_{1+1}^k) = \prod_{j=1}^{1} \widetilde{a}_j^k . \tag{4.5/20}$$

Because of eq. (4.3/15) and because of the fact that all vectors $\underline{\widetilde{d}}_j^k$ are scaled so that they have the Euclidean length 1 it follows

$$0 \leq \widetilde{a}_j^k \leq 1, \ j = 1, \ldots, 1 . \tag{4.5/21}$$

With respect to this inequality it follows from eq. (4.5/20) that the Gram determinant is always non-negative and that the adding of a further linearly independent vector $\widetilde{\underline{d}}_{1+1}^k$ for the building up the matrix $\widetilde{\underline{D}}_{1+1}^k$ causes the value of the determinant of the matrix $(\widetilde{\underline{D}}_{1+1}^k)^T\widetilde{\underline{D}}_{1+1}^k$ to decrease in general. When adding a further vector for the building up of the matrix $\widetilde{\underline{D}}_{1+1}^k$ the value of the Gram determinant remains unchanged if and only if this vector $\widetilde{\underline{d}}_{1+1}^k$ is orthogonal to all other vectors which have been selected previously.

The maximal value which is possible for the Gram determinant is 1. The Gram determinant has this value if and only if the scalar $\widetilde{a}_j^k$ has the value 1 for all j. This is true if and only if all selected vectors are mutually orthogonal. But in general we do not meet with this situation.

We obtain as large a value as possible of the Gram determinant of the matrix $\widetilde{\underline{D}}_1^k$ if we select the vectors for the building up of the matrix $\widetilde{\underline{D}}_1^k$ in such a way that the scalars $\widetilde{a}_j^k$ are maximal. But this is the processing when applying the criterion "maximal angle" for the selection of the vectors $\underline{d}_j^k$. This means that the selection procedure with this criterion guarantees that the value of the Gram determinant of the matrix $\widetilde{\underline{D}}_1^k$ becomes maximal. Referring to the condition number $\varkappa_H$ from eq. (4.5/7) this means that the denominator of this quotient is made as large as possible.

For the complete determination of the condition number $\varkappa_H$ it is necessary to calculate the numerator also. For that purpose we assume that we have selected 1 linearly independent vectors $\widetilde{\underline{d}}_1^k$, ..., $\widetilde{\underline{d}}_1^k$ for the building up of the matrix $\widetilde{\underline{D}}_1^k$. The value of the numerator of the condition number $\varkappa_H$ from eq. (4.5/7) belonging to these 1 selected vectors $\widetilde{\underline{d}}_j^k$ is denotated by $V_1$. Because of eq. (4.5/8) the value of $V_1$ is given by

$$V_1 = \prod_{j=1}^{1} \| (\widetilde{\underline{d}}_j^k)^T \widetilde{\underline{D}}_1^k \| = \sqrt{\prod_{j=1}^{1} (\widetilde{\underline{d}}_j^k)^T \widetilde{\underline{D}}_1^k (\widetilde{\underline{D}}_1^k)^T \widetilde{\underline{d}}_j^k} \quad . \tag{4.5/22}$$

By inspecting eq. (4.5/22) we see that the minimum value which is possible for $V_1$ is 1. This happens if and only if the vectors $\widetilde{\underline{d}}_1^k$, ..., $\widetilde{\underline{d}}_1^k$ are mutually orthogonal. Moreover the vectors $\widetilde{\underline{d}}_1^k$, ..., $\widetilde{\underline{d}}_1^k$ have the Euclidean length 1. Vectors with these two properties are called an orthonormalized set [1]. For such a set of orthonormalized vectors the Gram determinant has its maximum value 1 also, that means the condition number $\varkappa_H$ is 1. In all other cases it is greater than 1 because then the Gram determinant is smaller than 1 and the scalar $V_1$ is greater than 1.

Now we want to give an upper bound for the value of $V_1$ from eq. (4.5/22) by utilizing values which are calculated in the course of the selection procedure anyhow. By writing eq. (4.5/22) in a slightly modified way we obtain

$$V_1 = \sqrt{\prod_{j=1}^{1} \left\{ \sum_{i=1}^{1} [(\widetilde{\underline{d}}_j^k)^T \widetilde{\underline{d}}_i^k]^2 \right\}}$$

$$= \sqrt{\prod_{j=1}^{1} \left\{ 1 + \sum_{\substack{i=1 \\ i \neq j}}^{1} [(\widetilde{\underline{d}}_j^k)^T \widetilde{\underline{d}}_i^k]^2 \right\}} \quad . \tag{4.5/23}$$

In order to evaluate eq. (4.5/23) we need the scalar products $(\underline{\tilde{d}}_j^k)^T\underline{\tilde{d}}_i^k$ for $i \neq j$. If we do not want to calculate them we can give an upper bound for them. For that purpose assume that we have selected $1$ linearly independent vectors $\underline{\tilde{d}}_1^k$, ..., $\underline{\tilde{d}}_1^k$ for the building up of the matrix $\underline{\tilde{D}}_1^k$. In the course of this selection procedure we have determined the scalars $\tilde{a}_1^k$, ..., $\tilde{a}_{1-1}^k$. Therefore we can assume that these scalars are known.

For $j < i$ we have

$$[(\underline{\tilde{d}}_j^k)^T\underline{\tilde{d}}_i^k]^2 = \cos^2 \underline{\tilde{d}}_j^k, \underline{\tilde{d}}_i^k = 1 - \sin^2 \underline{\tilde{d}}_j^k, \underline{\tilde{d}}_i^k$$

$$= 1 - \|\underline{\tilde{d}}_{i_{\underline{d}_j^k}}^{k^\perp}\|^2 \ . \tag{4.5/24}$$

Because of the inequality (4.3/29) we can give an upper bound for this expression

$$[(\underline{\tilde{d}}_j^k)^T\underline{\tilde{d}}_i^k]^2 \leq 1 - \|\underline{\tilde{d}}_{i_{\underline{D}_{i-1}^k}}^{k^\perp}\|^2 = 1 - \tilde{a}_{i-1}^k \ . \tag{4.5/25}$$

As the scalar product from eq. (4.5/24) is commutative the inequality (4.5/25) is also valid for $i < j$. So we can give the following upper bound for $V_1$

$$V_1 \leq \sqrt{\prod_{j=1}^{1} \{1 + \sum_{i=1}^{1-1} (1-\tilde{a}_i^k)\}}$$

$$= \sqrt{\prod_{j=1}^{1} (1 - \sum_{i=1}^{1-1} \tilde{a}_i^k)} = [1 - \sum_{i=1}^{1-1} \tilde{a}_i^k]^{1/2} \ . \tag{4.5/26}$$

We recognize that it is possible to give an upper bound for $V_1$ if we know the values $\tilde{a}_1^k$, ..., $\tilde{a}_{1-1}^k$. But as the vectors $\underline{\tilde{d}}_j^k$ are selected successively we obtain the scalars $\tilde{a}_j^k$ also successively. This means that it is not necessary to store the values of the scalars $\tilde{a}_j^k$ in a digital computer but only the sum of them.

Utilizing the inequality (4.5/26) we can give an upper bound for the condition number $\varkappa_H$. It is given by

$$\varkappa_H((\underline{\tilde{D}}_1^k)^T\underline{\tilde{D}}_1^k) \leq \frac{[1 - \sum_{j=1}^{1-1} \tilde{a}_j^k]^{1/2}}{\prod_{j=1}^{1-1} \tilde{a}_j^k} \ . \tag{4.5/27}$$

With the inequality (4.5/27) we have an upper bound for the condition number $\varkappa_H$ which can easily be computed simultaneously with the selection procedure as we only need the scalars $\tilde{a}_j^k$ which are computed during the selection procedure anyhow.

By inspecting the inequality (4.5/27) we see that the criterion "maximal angle" for the selection of the vectors for the building up of the matrix $\underline{\tilde{D}}_1^k$ yields as small a value as possible of the upper bound for the condition number $\varkappa_H$ in each stage of the selection procedure because the numerator in the inequality (4.5/27) becomes minimal and the denominator maximal.

The upper bound for $\varkappa_H$ from the inequality (4.5/27) may be pretty inexact under certain circumstances because the inequality (4.5/25) may be very rough if the vectors $\tilde{\underline{d}}_j^k$ are not approximately mutually orthogonal. Therefore it can happen that we consider the matrix $(\underline{\tilde{D}}_l^k)^T\underline{\tilde{D}}_l^k$ as too ill-conditioned for an inversion if we examine this property by the aid of the upper bound from eq. (4.5/27) for the condition number. The utilization of this upper bound can cause the selection procedure of the vectors $\tilde{\underline{d}}_j^k$ for the building up of the matrix $\underline{\tilde{D}}_l^k$ to finish in order to avoid that the matrix to be inverted is too ill-conditioned. If we want to remove this shortcoming produced by the utilization of the upper bound for the condition number $\varkappa_H$ it is necessary to calculate the scalar $V_l$ exactly. In connexion with this it is to remark that the condition number itself is only a rough and approximate measure for the condition of the matrix to be inverted. Naturally the calculation of the scalar $V_l$ requires more numerical effort than the utilization of the relation (4.5/27) which turns out to be very satisfying when applying it to many practical problems which have been treated with the proposed new method for the solution of nonlinear least squares problems.

If we want to calculate the scalar $V_l$ so it is convenient to do this simultaneously with the selection procedure for the vectors $\tilde{\underline{d}}_j^k$. Assume we know $V_l$ and we add a further linearly independent vector $\tilde{\underline{d}}_{l+1}^k$ for the building up of the matrix $\underline{\tilde{D}}_{l+1}^k$. Then analogously to eq. (4.5/22) the scalar $V_{l+1}$ is given by

$$
\begin{aligned}
V_{l+1} &= \sqrt{\prod_{j=1}^{l+1} (\tilde{\underline{d}}_j^k)^T \underline{\tilde{D}}_{l+1}^k (\underline{\tilde{D}}_{l+1}^k)^T \tilde{\underline{d}}_j^k} \\
&= \sqrt{\prod_{j=1}^{l} [(\tilde{\underline{d}}_j^k)^T \underline{\tilde{D}}_l^k (\underline{\tilde{D}}_l^k)^T \tilde{\underline{d}}_j^k + (\tilde{\underline{d}}_j^k)^T \tilde{\underline{d}}_{l+1}^k (\tilde{\underline{d}}_{l+1}^k)^T \tilde{\underline{d}}_j^k]} \times \\
&\quad \times \sqrt{(\tilde{\underline{d}}_{l+1}^k)^T \underline{\tilde{D}}_l^k (\underline{\tilde{D}}_l^k)^T \tilde{\underline{d}}_{l+1}^k + 1} \quad .
\end{aligned}
\tag{4.5/28}
$$

We see that we have to know

$$
(\tilde{\underline{d}}_j^k)^T \underline{\tilde{D}}_l^k (\underline{\tilde{D}}_l^k)^T \tilde{\underline{d}}_j^k, \ j = 1, \ldots, l,
\tag{4.5/29}
$$

$$
(\tilde{\underline{d}}_j^k)^T \tilde{\underline{d}}_{l+1}^k (\tilde{\underline{d}}_{l+1}^k)^T \tilde{\underline{d}}_j^k, \ j = 1, \ldots, l,
\tag{4.5/30}
$$

and

$$
(\tilde{\underline{d}}_{l+1}^k)^T \underline{\tilde{D}}_l^k (\underline{\tilde{D}}_l^k)^T \tilde{\underline{d}}_{l+1}^k
\tag{4.5/31}
$$

in order to calculate $V_{l+1}$. The scalars from eq. (4.5/29) are necessary for the calculation of $V_l$ and are at our disposal. But these $l$ scalars have to be stored. The terms in the eqs. (4.5/30) and (4.5/31) are identical in the main so that only one calculation is necessary. But this calculation is not supplementary for the vector

$(\tilde{\underline{D}}_1^k)^T \tilde{\underline{d}}_{1+1}^k$ is needed for the computation of the vector $\tilde{\underline{u}}_1^k$ from eq. (4.3/17) anyhow. So it is clear that the supplementary numerical effort for the calculation of $V_{1+1}$ is not immense if we add a further linearly independent vector for the building up of the matrix $\tilde{\underline{D}}_{1+1}^k$. Therefore also an exact calculation of the condition number $\varkappa_H$ is possible and practicable.

By inspecting eq. (4.5/28) we see that $V_{1+1}$ increases in general when adding a further linearly independent vector for the building up of the matrix $\tilde{\underline{D}}_{1+1}^k$. $V_{1+1}$ remains unchanged if and only if the new vector $\tilde{d}_{1+1}^k$ is orthogonal to all already selected vectors. This is the only case in which the Gram determinant does not change. In all other cases the Gram determinant decreases so that the condition number $\varkappa_H$ increases when adding a further linearly independent vector for the building up of the matrix $\tilde{\underline{D}}_{1+1}^k$. This is the result corresponding to the inequality (4.5/6) but referring to the condition number $\varkappa_H$ when performing the selection of the vectors $\tilde{\underline{d}}_j^k$ in the system $\sim$.

If we consider a vector $\tilde{\underline{d}}_j^k$ and if for this vector the condition number $\varkappa_H$ exceeds a given upper bound which is necessary to guarantee a sufficiently well-conditioned matrix $(\tilde{\underline{D}}_1^k)^T \tilde{\underline{D}}_1^k$ this vector $\tilde{\underline{d}}_j^k$ is eliminated from the selection procedure of the vectors for the building up of the matrix $\tilde{\underline{D}}_1^k$. The vector $\tilde{\underline{d}}_j^k$ is eliminated from the selection procedure once and for all in the stage k of the iteration because the scalar $\varkappa_H$ belonging to this vector cannot decrease in a subsequent stage of the selection procedure.

It is intelligent to perform the examination of the condition number simultaneously with the selection procedure of the vectors for the building up of the matrix $\tilde{\underline{D}}_1^k$. Only those vectors $\tilde{\underline{d}}_j^k$ which cause a condition number $\varkappa_H$ which is less than a given upper bound take part in the selection procedure. The selection is performed according to one of the criteria presented in section 4.4, preferably according to the criterion "maximal change of the slope". The application of the criterion "maximal angle" for the selection causes the Gram determinant to become maximal. With that we only pay attention to the denominator of the condition number $\varkappa_H$. But because of eq. (4.5/28) we can also determine the numerator of the condition number $\varkappa_H$ simultaneously with the selection procedure so that it is possible to introduce a new criterion for the selection namely the criterion "minimal condition number $\varkappa_H$". Selecting a further vector $\tilde{\underline{d}}_j^k$ for the building up of the matrix $\tilde{\underline{D}}_1^k$ according to this criterion means that we add that vector next for which the condition number $\varkappa_H$ becomes minimal. Referring to the condition of the matrix to be inverted this selection procedure is the most suitable but in general it is not referring to the number of iteration steps necessary for the determination of the point $\underline{x}^*$.

## 4.6 Conditions for finishing the iteration

As it is well-known the necessary condition for a minimum of the function $Q(\underline{x})$ at the point $\underline{x}^k$ is the vanishing of the gradient of $Q(\underline{x})$ at $\underline{x} = \underline{x}^k$. Therefore it is obvious to take the absolute value of the gradient as a measure in order to decide whether the necessary condition for a minimum is satisfied or not. Because of the special structure (4.2/4) of the gradient in the case of nonlinear least squares problems it is possible to utilize another condition for finishing the iteration. This condition has some advantages in comparison with the examination of the absolute value of the gradient.

By inspecting eq. (4.2/4) we see that the gradient of the sum of squares $Q(\underline{x})$ vanishes if the error vector $\underline{e}(\underline{x}^k)$ is orthogonal to all column vectors of the matrix $\underline{D}^k$. Therefore it is convenient to examine the vanishing of the gradient by the aid of the angles between the error vector $\underline{e}(\underline{x}^k)$ and the column vectors $\underline{d}_1^k, \ldots, \underline{d}_n^k$ [15]. The examination of these angles is better suitable because the absolute values of the components of the gradient vector play no part. Moreover the calculation of these angles is easily done by means of values which are computed anyhow.

The n interesting angles $\beta_j^k$ are given for instance by

$$\cos\beta_j^k = \frac{(\underline{d}_j^k)^T \underline{e}(\underline{x}^k)}{\|\underline{d}_j^k\| \, \|\underline{e}(\underline{x}^k)\|} , \quad j = 1, \ldots, n \quad . \tag{4.6/1}$$

If $\cos\beta_j^k$ vanishes for all $j = 1, \ldots, n$ the necessary condition for a minimum is satisfied. Because of numerical inaccuracies we shall not succeed in satisfying this condition exactly. Instead of the theoretical value zero we demand that

$$\max_{1 \leq j \leq n} |\cos\beta_j^k| < \epsilon_1, \ \epsilon_1 > 0 \tag{4.6/2}$$

is valid. If the inequality (4.6/2) is satisfied for a given positive $\epsilon_1$ we consider the condition referring to the vanishing of the gradient as satisfied and we finish the iteration.

In order to evaluate eq. (4.6/1) with as small a numerical effort as possible we rewrite the relation (4.6/1) as follows

$$\cos\beta_j^k = -\frac{1}{2} \frac{\left.\frac{\partial Q}{\partial x_j}\right|_{\underline{x}^k}}{\sqrt{(\underline{d}_j^k)^T \underline{d}_j^k} \ \sqrt{\underline{e}(\underline{x}^k)^T \underline{e}(\underline{x}^k)}}$$

$$= -\frac{1}{2} \frac{\left.\frac{\partial Q}{\partial x_j}\right|_{\underline{x}^k}}{\sqrt{Q(\underline{x}^k)} \ \sqrt{(\underline{d}_j^k)^T \underline{d}_j^k}} \quad . \tag{4.6/3}$$

The values on the right-hand side of eq. (4.6/3) are all at our disposal. The gradient of $Q(\underline{x}^k)$ is needed for the determination of the direction vector. The sum of squares $Q(\underline{x}^k)$ is also computed because it is the function to be minimized and therefore it is of interest. Moreover it is needed for the one-dimensional search (see section 4.7). The Euclidean length of the vectors $\underline{d}_j^k$ has to be determined in order to perform the transformation (4.4.2/4). With that it is clear that the interesting angles can be calculated with a small supplementary numerical effort in comparison with the determination of the Euclidean length of the gradient. But this supplementary effort is justified because the angles $\beta_j^k$ are a well suitable measure for the examination of the necessary condition for a minimum.

Besides the condition (4.6/2) it is intelligent to introduce a further condition for finishing the iteration. This condition refers to the sum of squares Q. If the sum of squares has decreased so much that the value of $Q(\underline{x})$ is in the vicinity of zero that means if we have fitted the given fixed values $y_i$, $i = 1, \ldots, m$ nearly exactly it makes no sense to continue the iteration until the condition (4.6/2) is satisfied. Moreover in these cases there are numerical problems because of the division by $\sqrt{Q(\underline{x}^k)}$ in eq. (4.6/3) which is nearly zero.

Therefore we plan a further condition of the type

$$Q(\underline{x}^k) < \epsilon_2, \ \epsilon_2 > 0 \qquad (4.6/4)$$

where $\epsilon_2$ is a given positive value which will depend on the absolute values of the given fixed values $y_i$, on the number m of values to be fitted, and on the accuracy of the used digital computer.

We finish the iteration if either the condition (4.6/2) or the condition (4.6/4) is satisfied.

By evaluating eq. (4.6/3) we cannot only determine the angles $\beta_j^k$ between the error vector $\underline{e}(\underline{x}^k)$ and the column vectors of the matrix $\underline{D}^k$ but also which vector has to be selected first for the building up of the matrix $\underline{D}_1^{k+1}$. As it is well-known we have to select that vector first for which the expression (4.3/2) becomes minimal. By utilizing the relation (4.6/1) we obtain from eq. (4.3/2)

$$- 2 \, \underline{e}(\underline{x}^k)^T \underline{d}_j^k ((\underline{d}_j^k)^T \underline{d}_j^k)^{-1} (\underline{d}_j^k)^T \underline{e}(\underline{x}^k)$$

$$= - 2 \|\underline{d}_j^k\|^2 \ \|\underline{e}(\underline{x}^k)\|^2 \ \|\underline{d}_j^k\|^{-2} \cos^2 \beta_j^k$$

$$= - 2 \, Q(\underline{x}^k) \cos^2 \beta_j^k \ . \qquad (4.6/5)$$

This expression becomes minimal for that vector $\underline{d}_j^k$ for which $\cos^2 \beta_j^k$ becomes maximal. But this is the same vector which yields the maximum of $|\cos \beta_j^k|$. If we evaluate the condition (4.6/2) in order to examine the necessary condition for finishing the iteration we simultaneously obtain the information which vector has to be selected first for the building up of the matrix $\underline{D}_1^{k+1}$ if a continuation of the iteration is necessary at all.

When applying the two above described conditions (4.6/2) and (4.6/4) for finishing the iteration the principal components of the new method for the solution of nonlinear least squares problems are represented in the logic diagram 11. Those components of the method which are shown in details in the logic diagrams 6, 7, 8, and 9 are reproduced in a shortened manner.

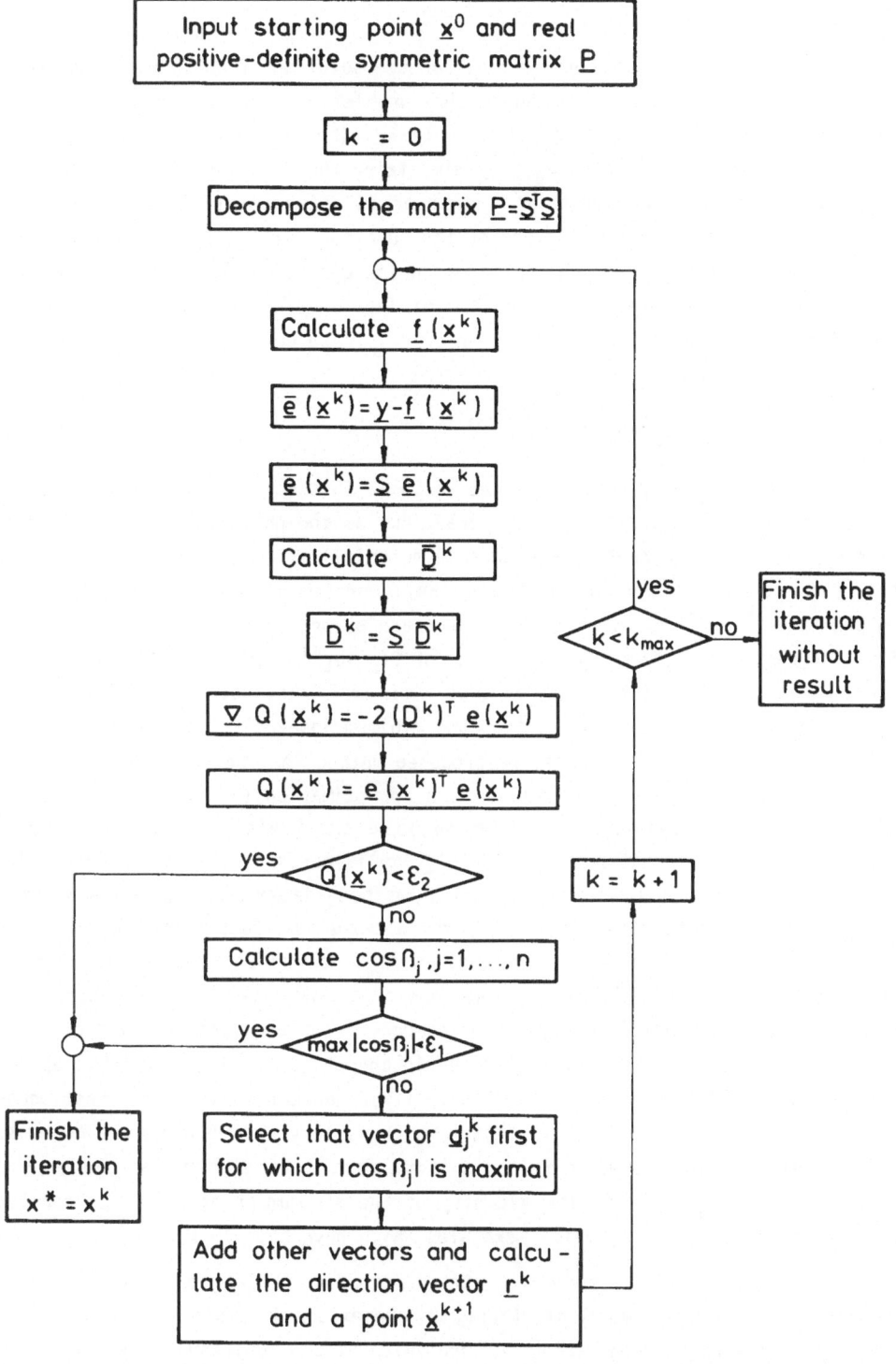

Logic diagram 11: Examination of the conditions for finishing the iteration and selection of the first vector for the building up of the matrix $\underline{D}_1^k$.

## 4.7   A procedure for the one-dimensional search

In connexion with the proposed new method for the solution of nonlinear least squares problems we have not yet considered the problem of the determination of the step-length factor $\alpha_{min}$. Returning to the logic diagram 6 we see that we have to determine the step-length factor $\alpha_{min}$ by minimizing the function $Q(\underline{x}^k+\alpha\underline{r}^k)$ after having computed the direction vector $\underline{r}^k$. First of all it is guaranteed because of the stability of the method that the slope of the function $Q(\underline{x}^k+\alpha\underline{r}^k)$ at the point $\underline{x}^k$ is negative, that means the function decreases in the direction of positive values of $\alpha$. So it is clear that we have to look for the minimum of $Q(\underline{x}^k+\alpha\underline{r}^k)$ only for positive values of $\alpha$. Therefore it is sufficient in principle to determine such a value $\alpha > 0$ for which

$$Q(\underline{x}^k+\alpha\underline{r}^k) < Q(\underline{x}^k) \qquad\qquad (4.7/1)$$

holds, for instance a very small positive value $\alpha$. This value only causes a small decrease of the sum of squares $Q(\underline{x})$ in general. But as the determination of the direction vector $\underline{r}^k$ requires a pretty considerable numerical effort compared with all other necessary tasks we are not interested in any decrease of the sum of squares $Q(\underline{x})$ along the line $\underline{x}^k+\alpha\underline{r}^k$ but in as great a decrease as possible. But this means that we must be interested in determining a minimum of $Q(\underline{x}^k+\alpha\underline{r}^k)$, if possible the absolute minimum. First it is not clear whether such a minimum exists at all. As $Q(\underline{x})$ only has non-negative values it could happen that the minimum appears for $\alpha \to \infty$. But such a behaviour is not to expect because the proposed method for the solution of nonlinear least squares problems is also based on the approximation (3.2/1) for the function $\underline{f}(\underline{x})$ and this approximation is only valid in a small vicinity of the point $\underline{x}^k$. Therefore we can expect that the function $Q(\underline{x}^k+\alpha\underline{r}^k)$ increases when performing a large step in the direction of the vector $\underline{r}^k$ which has been determined on the basis of the approximation of the function $\underline{f}(\underline{x})$. Moreover we have the fact that the step-length factor $\alpha_{min}$ has the value 1 in the case of a linear least squares problem (see chapter 5, especially eq. (5/10)). In the case of a linear least squares problem the function $\underline{f}(\underline{x})$ only depends linearly on the components of the parameter vector $\underline{x}$ to be determined. This special problem is treated in chapter 5. In this case the approximation (3.2/1) is identical with the function $\underline{f}(\underline{x})$. Hence for nonlinear least squares problems we can expect a step-length factor which is less than 1 if we are "far away" from the minimum point $\underline{x}^*$ we are looking for. On the other hand we can expect a step-length factor $\alpha_{min} \approx 1$ in the vicinity of the minimum point $\underline{x}^*$. These considerations are corroberated by numerical examples which have been performed.

It remains the problem to determine the step-length factor $\alpha_{min}$ relatively exactly but without considerable numerical effort. In connexion with this we must pay attention to the fact that the calculation of one value of the function $Q(\underline{x}^k+\alpha\underline{r}^k)$ for a given value $\alpha$ can be very expensive depending on the special function $\underline{f}(\underline{x})$ under consideration. Therefore we are interested in obtaining an optimum of information about the possible value of $\alpha_{min}$ on the basis of the already known properties of the function $Q(\underline{x})$ - i. e. the value of $Q(\underline{x}^k)$ and the slope of the function $Q(\underline{x}^k + \alpha \underline{r}^k)$ for $\alpha = 0$. In general these informations are not sufficient for the determination of the step-length factor. So we have to perform further function evaluations whose number should be as small as possible. A procedure which is appropriate for this traces back to Davidon [7]. The principal components of his procedure - modified for the application to nonlinear least problems - are represented in the sequel. Besides this procedure there is a lot of others which are all more or less efficient (see for instance [7, 16, 17]).

The procedure based on Davidon's idea consists of 3 phases

Phase 1: Determination of a step-length factor $\alpha_s$ to begin with

Phase 2: Determination of an interval $(\alpha', \alpha'')$

Phase 3: Cubic interpolation in the interval $(\alpha', \alpha'')$ and approximation of the minimum.

The processing in the 3 phases is described in the sequel.

Phase 1:
In the case of nonlinear least squares problems the step-length factor $\alpha_s$ to begin with is set to the value 1 on the analogy of the linear least squares problem (compare eq. (5/10)). The value of $Q(\underline{x}^k + \underline{r}^k)$ is calculated. If

$$Q(\underline{x}^k + \alpha_s \underline{r}^k) > \mu Q(\underline{x}^k), \quad \mu \gg 1 \qquad (4.7/2)$$

holds the value of $\alpha_s$ is decreased. This decrease is performed several times depending on the inequality (4.7/2). If this inequality is not satisfied the value of $\alpha_s$ is accepted. Numerical investigations have shown that the choice $\mu = 1000$ is suitable.

Phase 2:
In this phase we calculate the values of the function $Q(\underline{x}^k + \alpha \underline{r}^k)$ for the following values of $\alpha$

$$\alpha = \alpha_s, \ \alpha = 2\alpha_s, \ ..., \ \alpha = 2^r \alpha_s, \ r = 0, 1, 2, ... \quad . \qquad (4.7/3)$$

We double $\alpha$ until the relation

$$Q(\underline{x}^k + \alpha_1 \underline{r}^k) > Q(\underline{x}^k + \alpha_2 \underline{r}^k) < Q(\underline{x}^k + \alpha_3 \underline{r}^k) \qquad (4.7/4)$$

with

$$\alpha_3 = 2^r \alpha_s$$
$$\alpha_2 = 2^{r-1} \alpha_s$$
$$\alpha_1 = 2^{r-2} \alpha_s \qquad (4.7/5)$$

holds. (In the case $r = 1$ we set $2^{r-2}$ equal to zero.)

The relation (4.7/4) means that we increase the step-length factor $\alpha$ so far until we obtain a value for the function $Q(\underline{x}^k + \alpha \underline{r}^k)$ which is greater than the preceding one. This situation is visualized in figure 4/1.

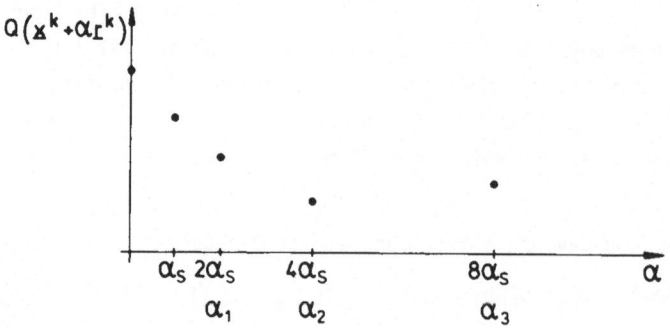

Fig. 4/1: Doubling of the step-length factor $\alpha_s$ in the phase 2 of the one-dimensional search.

Now the slope of the function $Q(\underline{x}^k + \alpha \underline{r}^k)$ for $\alpha = \alpha_2$ is calculated. If the slope is positive we continue the one-dimensional search in the interval

$$(\alpha', \alpha'') = (\alpha_1, \alpha_2) \ . \tag{4.7/6}$$

If the slope is negative we look for the value $\alpha_{min}$ in the interval

$$(\alpha', \alpha'') = (\alpha_2, \alpha_3) \ . \tag{4.7/7}$$

If the slope is zero the necessary condition for a minimum is satisfied. In this case we finish the one-dimensional search that means phase 3 of the procedure is not performed.

The different possibilities in phase 2 are represented in figure 4/2.

It can happen that

$$Q(\underline{x}^k + \underline{r}^k) > Q(\underline{x}^k) \tag{4.7/8}$$

holds. This case comes forward sometimes if the starting point $\underline{x}^0$ is selected so that it is "far away" from the minimum point $\underline{x}^*$. Here we set

$$(\alpha', \alpha'') = (0, 1) \ . \tag{4.7/9}$$

Phase 3:

After the interval $(\alpha', \alpha'')$ has been determined for the further search the function $Q(\underline{x}^k + \alpha \underline{r}^k)$ is approximated by a cubic parabola in the interval $(\alpha', \alpha'')$. The parameters of this parabola are given by the known values

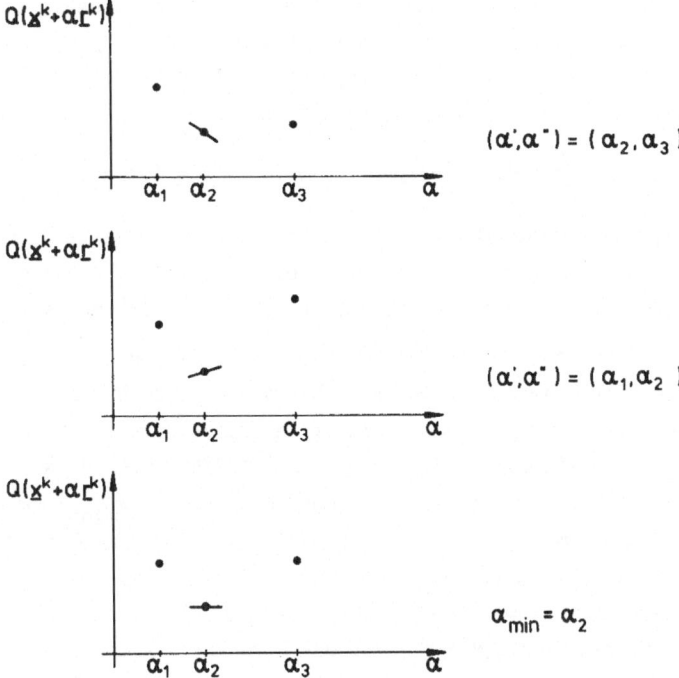

Fig. 4/2: Determination of the interval $(\alpha',\alpha'')$.

$$Q(\underline{x}^k+\alpha'\underline{r}^k)$$

$$Q(\underline{x}^k+\alpha''\underline{r}^k) \tag{4.7/10}$$

$$\frac{dQ}{d\alpha}\Big|_{\alpha'} \text{ respectively } \frac{dQ}{d\alpha}\Big|_{\alpha''}$$

and the value

$$\frac{dQ}{d\alpha}\Big|_{\bar{\alpha}''} \text{ respectively } \frac{dQ}{d\alpha}\Big|_{\alpha'} \tag{4.7/11}$$

which has to be calculated supplementarily. For this approximating parabola we can calculate the value $\alpha*$ for which it has a minimum. The value $\alpha*$ depends on the values from eqs. (4.7/10) and (4.7/11). For the value $\alpha*$ we evaluate $Q(\underline{x}^k+\alpha\underline{r}^k)$. If

$$Q(\underline{x}^k+\alpha'\underline{r}^k) > Q(\underline{x}^k+\alpha*\underline{r}^k) < Q(\underline{x}^k+\alpha''\underline{r}^k) \tag{4.7/12}$$

holds we set

$$\alpha_{min} = \alpha* \quad . \tag{4.7/13}$$

By eq. (4.7/13) an approximate value for $\alpha_{min}$ is given. But this value is accepted in order to keep the numerical effort for the one-dimensional search small. If the inequality relation (4.7/12) does not hold the slope of the function $Q(\underline{x}^k + \alpha \underline{r}^k)$ is calculated for $\alpha = \alpha^*$. Depending on the fact whether it is positive or negative we set

$$(\alpha', \alpha'') = (\alpha', \alpha^*) \text{ respectively } (\alpha', \alpha'') = (\alpha^*, \alpha'') \ . \qquad (4.7/14)$$

Now a further cubic approximating parabola is determined. The necessary values for the determination of its four parameters are all at our disposal. The phase 3 of this search procedure is repeated until the inequality relation (4.7/12) holds.

Referring to the proposed new method for the solution of nonlinear least squares problems the experience of many examples shows that this above described one-dimensional search procedure in general requires no doubling of the step-length factor $\alpha_s$ to begin with so that the interval $(\alpha', \alpha'')$ is given by eq. (4.7/9). For the most part one passage of the phase 3 is sufficient in order to obtain the value $\alpha_{min}$ from eq. (4.7/13). This means that in most cases only two evaluations of the function $Q(\underline{x})$ - namely $Q(\underline{x}^k + \underline{r}^k)$ and $Q(\underline{x}^k + \alpha \underline{r}^k)$ - and one calculation of the slope of the function $Q(\underline{x}^k + \alpha \underline{r}^k)$ for $\alpha = 1$ have to be performed. The value of $Q(\underline{x}^k)$ is known just so the value of the slope of the function $Q(\underline{x}^k + \alpha \underline{r}^k)$ for $\alpha = 0$ for the latter value can easily be calculated without considerable numerical effort in the course of the selection procedure of the vectors $\underline{d}_j^k$ for the building up of the matrix $\underline{D}_1^k$. This is true in particular if we utilize the criterion "maximal change of the slope" for the selection of the vectors $\underline{d}_j^k$.

Comprehensively we can state that the determination of an approximate value for $\alpha_{min}$ can easily be performed by the aid of the above described procedure for the one-dimensional search. The numerical effort for it is very small.

References:

[1] Young, D.M.; Gregory, R.T.:
A Survey of Numerical Mathematics
Vol. II
Reading, Massachusetts: Addison-Wesley Publishing Company, Inc. (1973).

[2] Marquardt, D.W.:
Generalized Inverses, Ridge Regression, Biased Linear Estimation and Nonlinear Estimation
Technometrics 3, 591-612 (1970).

[3] Faddejew, D.K.; Faddejewa, W.N.:
Numerische Methoden der linearen Algebra
4. Auflage
München und Wien: R. Oldenbourg Verlag (1976).

[4] Zurmühl, R.:
Matrizen
4. Auflage
Berlin, Heidelberg, New York: Springer-Verlag (1964).

[5] Fletcher, R.:
Generalized Inverse Method for the Best Least Squares Solutions of Systems of Nonlinear Equations
The Computer Journal 10, 392-399 (1967).

[6] Jennrich, R.I.; Sampson, P.F.:
Application of Stepwise Regression to Nonlinear Estimation
Technometrics 1, 63-72 (1968).

[7] Pierre, D.A.:
Optimization Theory with Applications
New York: J. Wiley & Sons, Inc. (1969).

[8] Rosen, J.B.:
The Gradient Projection Method for Nonlinear Programming
Part I. Linear Constraints
SIAM J. Appl. Math. 8, 181-217 (1960).

[9] Marquardt, D.W.:
An Algorithm for Least-Squares Estimation of Nonlinear Parameters
SIAM J. Appl. Math. 2, 431-441 (1963).

[10] Bard, Y.:
Comparison of Gradient Methods for the Solution of Nonlinear Parameter Estimation Problems
SIAM J. Numer. Anal. 1, 157-186 (1970).

[11] Stoer, J.:
Einführung in die Numerische Mathematik I
Heidelberger Taschenbücher
Berlin, Heidelberg, New York: Springer-Verlag (1972).

[12] Young, D.M.; Gregory, R.T.:
A Survey of Numerical Mathematics
Vol. I
Reading, Massachusetts: Addison-Wesley Publishing Company, Inc. (1972).

[13] Hanson, R.J.; Lawson, Ch.L.:
Extensions and Applications of the Householder Algorithm for Solving Linear Least Squares Problems
Math. Comp. 23, 787-812 (1969).

[14] Luenberger, D.G.:
Optimization by Vector Space Methods
New York: J. Wiley & Sons, Inc. (1969).

[15] Dennis, J.E.:
Nonlinear Least Squares and Equations
Proc. of the Conference on the State of the Art in Numerical Analysis, April 12th - 15th, 1976. Edited by D. Jacobs.
New York and London: Academic Press (1977).

[16] Bard, Y.:
Nonlinear Parameter Estimation
New York and London: Academic Press (1974).

[17] Himmelblau, D.M.:
Applied Nonlinear Programming
New York: McGraw-Hill, Inc. (1972).

5. APPLICATION OF THE NEW METHOD FOR THE SOLUTION OF THE LINEAR LEAST SQUARES PRO-
   BLEM

A special case of nonlinear least squares problems is the linear one, i. e. that problem for which the functions $f_i$, i = 1, ..., m depend linearly on the parameter vec-
tor $\underline{x}$ which is to be determined. Such a linear function $f_i(\underline{x})$ can be represented by

$$f_i(\underline{x}) = \underline{a}_i^T \underline{x}. \qquad (5/1)$$

If we gather up all functions $f_i(\underline{x})$, i = 1, ..., m in the vector $\underline{f}(\underline{x})$ and all row
vectors $\underline{a}_i^T$ in the matrix $\underline{A}$, we can write

$$\underline{f}(\underline{x}) = \underline{A}\ \underline{x} \qquad (5/2)$$

whereby $\underline{A}$ is a mxn-matrix. Solving the linear least squares problem means to find the
vector $\underline{x}^*$ for which

$$Q(\underline{x}) = (\underline{y} - \underline{A}\ \underline{x})^T \underline{P}(\underline{y} - \underline{A}\ \underline{x}) \qquad (5/3)$$

is minimal. $Q(\underline{x})$ can be expressed with the help of the decomposition $\underline{P} = \underline{S}^T \underline{S}$ and the
error vector

$$\underline{e}(\underline{x}) = \underline{S}(\underline{y} - \underline{A}\ \underline{x}) \qquad (5/4)$$

by

$$Q(\underline{x}) = (\underline{e}(\underline{x}))^T \underline{e}(\underline{x}) = \|\underline{e}(\underline{x})\|^2. \qquad (5/5)$$

Now we have the task to determine that vector $\underline{x}^*$ with the help of the proposed method
for the solution of nonlinear least squares problems for which $Q(\underline{x})$ from eq. (5/5)
becomes minimal. For that purpose we assume that we know a vector $\underline{x}^k$ from which we
start the determination of the point $\underline{x}^*$. For the determination of the direction vector $\underline{r}^k$ it is necessary to calculate the matrix $\underline{D}^k$. Because of the eqs. (A/5) and
(3.1/11) we obtain for it

$$\underline{D}^k = \underline{S}\ \overline{\underline{D}}^k = \underline{S}\ \underline{A}. \qquad (5/6)$$

Eq. (5/6) means that the matrix $\underline{D}^k$ is independent on the actually reached stage of
the iteration. Therefore the matrix $\underline{D}^k$ has to be calculated only once during the
whole iteration. Therefore we write

$$\underline{D}^k = \underline{D}. \qquad (5/7)$$

With this notation we obtain for the error vector $\underline{e}(\underline{x}^k)$ in the stage k of the iteration

$$\underline{e}(\underline{x}^k) = \underline{S}\ \underline{y} - \underline{D}\ \underline{x}^k. \qquad (5/8)$$

The next point $\underline{x}^{k+1}$ of the iteration is given by

$$\underline{x}^{k+1} = \underline{x}^k + \alpha_{min}\underline{r}^k \tag{5/9}$$

whereby the vector $\underline{r}^k$ is given by eq. (4.2/33). The value $\alpha_{min}$ need not be determined by a numerical method for the one-dimensional search; rather it is given a priori. For linear least squares problems we have

$$\alpha_{min} = 1. \tag{5/10}$$

Eq. (5/10) is always valid and that independently on the number of column vectors selected for the building up of the matrix $\underline{D}_1^k$. In the case that the column vectors of the matrix $\underline{D}$ are linearly independent this result is well known because we have the Gauss-Newton method.

In the cases in which the column vectors of the matrix $\underline{D}$ are not linearly independent respectively not all column vectors of the matrix $\underline{D}$ are selected for the building up of the matrix $\underline{D}_1^k$, the relation (5/10) has to be proved. This is an easy task if we take notice of the fact that the relation

$$(\underline{r}^k)^T \underline{\nabla} Q(\underline{x}^k + \alpha_{min}\underline{r}^k) = 0 \tag{5/11}$$

is valid at the point $\underline{x}^k + \alpha_{min}\underline{r}^k$. If we set in the gradient of Q at the point $\underline{x}^k + \alpha_{min}\underline{r}^k$ - which we find by evaluation of the relation (4.2/4) with regard to eq. (5/8) - we obtain

$$\alpha_{min}(\underline{r}^k)^T\underline{D}^T\underline{D}\ \underline{r}^k - (\underline{r}^k)^T\underline{D}^T\underline{S}\ \underline{y} + (\underline{r}^k)^T\underline{D}^T\underline{D}\ \underline{x}^k = 0. \tag{5/12}$$

From eq. (5/12) we find

$$\alpha_{min} = \frac{(\underline{r}^k)^T\underline{D}^T\underline{S}\ \underline{y} - (\underline{r}^k)^T\underline{D}^T\underline{D}\ \underline{x}^k}{(\underline{r}^k)^T\underline{D}^T\underline{D}\ \underline{r}^k} . \tag{5/13}$$

The denominator in eq. (5/13) does not vanish. Because of eq. (4.2/10) we have the result

$$\underline{D}\ \underline{r}^k = \underline{e}(\underline{x}^k)" \text{ with } \underline{e}(\underline{x}^k)" \in R(\underline{D}_1^k). \tag{5/14}$$

In eq. (5/14) $R(\underline{D}_1^k)$ means the subspace spanned by the selected column vectors. But as the first vector for the building up of the matrix $\underline{D}_1^k$ is selected so that the absolute value of the component of the error vector $\underline{e}(\underline{x}^k)$ in the space spanned by this selected column vector becomes as great as possible and as the inequality (4.4.1/8) is valid, we obtain

$$(\underline{r}^k)^T\underline{D}^T\underline{D}\ \underline{r}^k > 0. \tag{5/15}$$

To show the validity of eq. (5/10) we must utilize the direction vector $\underline{r}^k$ - as given by the relation (4.2/33) - in eq. (5/13). After a short conversion we obtain the pretended result. With that we have proved that the determination of the step-length factor is unnecessary if we are solving the linear least squares problem. This result is independent on the number of selected vectors $\underline{d}_j^k$ for the building up of the matrix $\underline{D}_1^k$. It remains to show that a minimum really exists for $\alpha_{min} = 1$. This can be seen by considering the second derivative

$$\frac{d^2 Q(\underline{x}^k + \alpha \underline{r}^k)}{d\alpha^2} = (\underline{r}^k)^T \underline{D}^T \underline{D} \; \underline{r}^k. \tag{5/16}$$

The second derivative is positive because of the inequality (5/15). Therefore we really have a minimum.

Next we shall prove a relation which is very important in the sequel. If in the stage k of the iteration we select l vectors for the building up of the matrix $\underline{D}_1^k$, then

$$(\underline{D}_1^k)^T \underline{e}(\underline{x}^{k+1}) = \underline{0} \tag{5/17}$$

is valid. Eq. (5/17) means that the error vector $\underline{e}(\underline{x}^{k+1})$ in the stage k + 1 of the iteration is orthogonal to the column vectors $\underline{d}_1, \ldots, \underline{d}_l$ selected in the stage k of the iteration. Because of eq. (4.2/4) this means that the components of the gradient vector - corresponding to the l selected vectors $\underline{d}_1^k, \ldots, \underline{d}_l^k$ - vanish in the stage k + 1 of the iteration.

For the proof of the validity of eq. (5/17) we write down $\underline{e}(\underline{x}^{k+1})$. It is given by

$$\underline{e}(\underline{x}^{k+1}) = \underline{S} \; \underline{y} - \underline{D} \; \underline{x}^{k+1} = \underline{S} \; \underline{y} - \underline{D}(\underline{x}^k + \underline{r}^k)$$

$$= \underline{S} \; \underline{y} - \underline{D} \; \underline{x}^k - \underline{D} \; \overline{\underline{R}}^k \underline{S} \; \underline{y} + \underline{D} \; \overline{\underline{R}}^k \underline{D} \; \underline{x}^k$$

$$= (\underline{I} - \underline{D} \; \overline{\underline{R}}^k)\underline{S} \; \underline{y} - (\underline{I} - \underline{D} \; \overline{\underline{R}}^k)\underline{D} \; \underline{x}^k . \tag{5/18}$$

The matrix $\overline{\underline{R}}^k$ is given by eq. (4.2/33). If we now form the product (5/17), we obtain

$$(\underline{D}_1^k)^T \underline{e}(\underline{x}^{k+1}) = [(\underline{D}_1^k)^T - (\underline{D}_1^k)^T \underline{D} \; \overline{\underline{R}}^k]\underline{S} \; \underline{y} - [(\underline{D}_1^k)^T - (\underline{D}_1^k)^T \underline{D} \; \overline{\underline{R}}^k]\underline{D} \; \underline{x}^k. \tag{5/19}$$

The right-hand side of eq. (5/19) vanishes because of eq. (4.2/30), that means on account of the fact that the matrix $\underline{D} \; \overline{\underline{R}}^k$ is an orthogonal projector which takes a vector into the column space of the matrix $\underline{D}_1^k$. Eq. (5/17) means that the gradient $\underline{\nabla}Q(\underline{x}^{k+1})$ vanishes referring to the components of the vector $\underline{x}$ modified in the stage k of the iteration. So it is guaranteed that certainly none of the vectors selected in the stage k of the iteration will be selected as first in the stage k + 1 of the iteration because the expression (4.3/2) which is responsible for the selection of the first vector vanishes. So it is clear that in the stage k + 1 of the iteration at most one vector will be selected which causes one component of the vector $\underline{x}$ - not modified in the stage k of the iteration - to be modified now.

Furthermore we recognize by eq. (5/17) that we find the minimum in one step if the column vectors of the matrix $\underline{D}$ are linearly independent. This is the well-known result of the Gauss-Newton method.

Moreover we reach the minimum in exactly one step if we are able to determine the theoretical rank r of the matrix $\underline{D}$ also with the help of a digital computer. (r is the maximal number of linearly independent column vectors of the matrix $\underline{D}$ with r $\leq$ n.) Then the relation (4.2/18) is valid for the linearly dependent column vectors. From it together with eq. (5/17) we obtain

$$\underline{D}^T\underline{e}(\underline{x}^{k+1}) = \underline{0}. \tag{5/20}$$

So it is proved that the proposed method yields the solution of the linear least squares problem in one step if we are able to determine the theoretical rank r of the matrix $\underline{D}$.

This result means that the proposed method is quadratically convergent in one step. The universal gradient methods presented in section 3.1 are in general quadratically convergent in n steps, for instance the Davidon-Fletcher-Powell algorithm. A method which is quadratically convergent in n steps solves the linear least squares problem in at most n steps but not in one. Concerning the solution of nonlinear least squares problems this fact means that at least in the vicinity of the minimum $\underline{x}^*$ the universal gradient methods are more disadvantageous than the method proposed here.

So far, we have only considered the case that we are able to determine the theoretical rank also numerically. In the other cases, i. e. if we have to assign a rank to the matrix $\underline{D}$ because of numerical reasons the minimum will not be reached in one step. Because of the guaranteed stability of the method the sum of squares Q decreases however from one stage of the iteration to another. This decrease of the sum of squares Q can easily be calculated. For that purpose we assume that we select l column vectors for the building up of the matrix $\underline{D}_l^k$ in the stage k of the iteration. Utilizing the notation $\underline{e}(\underline{x}^k)"_{\underline{D}_l^k}$ from eq. (4.4.1/3) we obtain

$$Q(\underline{x}^{k+1}) = Q(\underline{x}^k) - \|\underline{e}(\underline{x}^k)"_{\underline{D}_l^k}\|^2 . \tag{5/21}$$

Eq. (5/21) can be derived by applying eq. (5/5) in connexion with eq. (5/9) for calculating $Q(\underline{x}^k)$ and $Q(\underline{x}^{k+1})$. Assume that we only select one column vector for the building up of the matrix $\underline{D}_l^k$ in the stage k of the iteration, so it is certainly convenient to demand that we select that vector which causes as great a decrease as possible of the sum of squares. But this means that it is necessary to make $\|\underline{e}(\underline{x}^k)"_{\underline{D}_l^k}\|^2$ as great as possible. We reach our goal if we select the first vector so that the expression (4.3/2) becomes as small as possible. This means that the demand for as small a value as possible of the right-hand side of eq. (4.3/2) in the case of nonlinear least squares problems causes as great a decrease as possible of the sum of squares in the case of the linear least squares problem.

By eq. (5/21) we recognize moreover that it is necessary to make $||\underline{e}(\underline{x}^k)"_{\underline{D}_1^k}||^2$ as great as possible to achieve as great a decrease as possible of the sum of squares in one stage of the iteration. From eq. (4.4.1/8) we see that an increase of $||\underline{e}(\underline{x}^k)"_{\underline{D}_1^k}||$ is possible by adding further linearly independent vectors for the building up of the matrix $\underline{D}_1^k$. The change is just maximal if we select that vector as next for which the expression given by eq. (4.4.1/10) becomes maximal. This processing corresponds to the selection of the vectors according to the criterion "maximal change of the slope". This means, if we apply the proposed method for the solution of the linear least squares problem and if we select the vectors $\underline{d}_j^k$ for the building up of the matrix $\underline{D}_1^k$ according to the criterion "maximal change of slope", we obtain a maximal decrease of the sum of squares by adding one further vector. So the criterion for the selection of the vectors $\underline{d}_j^k$ given in the case of the solution of nonlinear least squares problems is well founded when applying to the solution of the linear least squares problem.

Considering eq. (5/21) we can derive a condition how many linearly independent vectors $\underline{d}_j^k$ should be selected for the building up of the matrix $\underline{D}_1^k$. For instance we can demand that

$$Q(\underline{x}^{k+1}) \leq b^k Q(\underline{x}^k) \text{ with } 0 < b^k < 1 \qquad (5/22)$$

is valid. From this demand we obtain the condition

$$||\underline{e}(\underline{x}^k)"_{\underline{D}_1^k}||^2 \geq (1-b^k)Q(\underline{x}^k). \qquad (5/23)$$

The inequality (5/23) means that we select further linearly independent vectors for the building up of the matrix $\underline{D}_1^k$ until the inequality (5/23) is satisfied. The evaluation of the inequality is not expensive because on account of the relation (4.4.1/9) the examination of the validity of the inequality takes place simultaneously with the selection of the vectors $\underline{d}_j^k$ for the building up of the matrix $\underline{D}_1^k$. Naturally it is possible that the inequality (5/23) cannot be satisfied by selecting all numerically determined linearly independent vectors. This simply means that the scalar $b^k$ was chosen too small. Therefore the value $b^k$ has to be increased in the stage $k + 1$ of the iteration. In principle many different strategies for the modification of the scalar $b^k$ from one stage of the iteration to another are imaginable. It is obvious to extrapolate the new value $b^k$ from the values of the sum of squares achieved in the previous stages of the iteration. Depending on the number of values of the sum of squares which are at our disposal from the previous stages of the iteration it is possible to extrapolate linearly, quadratically, and so on.

The criterion given by the inequality (5/23) for the fixation of the number 1 of the linearly independent vectors which are selected for the building up of the matrix $\underline{D}_1^k$ is naturally also applicable in the case of the solution of nonlinear least squares problems. Then however, it is not guaranteed that the decrease of the sum of squares Q given by eq. (5/21) is really achieved.

# 6. THE PROBLEM OF THE CHOICE OF A STARTING POINT $\underline{x}^0$

In order to start the proposed method for the solution of nonlinear least squares problems it is necessary to input a starting point $\underline{x}^0$ (compare logic diagram 6). Certainly, it depends on the choice of this starting point $\underline{x}^0$ whether and in case in how many iteration steps the point $\underline{x}^*$ we are looking for will be found. In the special case of the linear least squares problem with the possibility of determining the theoretical rank of the matrix $\underline{D}^k$ on a digital computer the starting point $\underline{x}^0$ plays no part because the point $\underline{x}^*$ we are looking for is always found in one step independently on the actually chosen special starting point $\underline{x}^0$. If in the case of nonlinear least squares problems a point $\underline{x}^k$ is reached during the iteration for which the function $\underline{f}(\underline{x})$ is sufficiently exactly described by its linear approximation and if we are able to determine the theoretical rank of the matrix $\underline{D}^k$, we obtain a point $\underline{x}^{k+1}$ for which the gradient of the sum of squares vanishes in the next stage of the iteration.

This point can be considered as a solution of the problem because it satisfies the necessary condition (3.1/2). If we do not reach a point for which the function $\underline{f}(\underline{x})$ is sufficiently exactly described by its linear approximation (3.2/1), so we can however achieve a decrease of the sum of squares because of the guaranteed stability of the method. So we approach the point $\underline{x}^*$ step by step. Qualitatively, it is clear that the search for the point $\underline{x}^*$ requires more iteration steps in general if the starting point $\underline{x}^0$ is "far away" from the yet unknown point $\underline{x}^*$. But it happens also that a starting point $\underline{x}^0$ in the vicinity of $\underline{x}^*$ requires more iteration steps than other starting points "far away" from $\underline{x}^*$. This means that the number of iteration steps increases if we choose a "bad" starting point $\underline{x}^0$. Therefore it is useful to look for as good a starting point as possible, for instance by exploiting supplementary informations. Such supplementary informations can often be obtained from the physics of the problem under consideration especially in the case of nonlinear least squares problems in technical applications. So it often happens that the order of magnitude of some parameters is known but not their exact values. From this knowledge we can obtain an essential indication for the choice of some components of the starting point $\underline{x}^0$. But from this knowledge great benefit can only be derived if the approximately known components of the starting vector $\underline{x}^0$ appear explicitly as parameters in the function $\underline{f}(\underline{x})$. This state of affairs is to be illustrated by an example from control theory.

Assume we have measured the step response of a linear time-invariant dynamic system with one input u and one output y for some points of time $t_i$. The dynamic behaviour of such a system can be described by a transfer function $\hat{G}(s)$ if we consider the case that all initial values of the system are zero.

For the solution of many problems in control theory it is necessary to know the transfer function $\hat{G}(s)$. Therefore the parameters of the transfer function

$$\hat{G}(s) = \frac{b_0 + b_1 s + \ldots + b_p s^p}{a_0 + a_1 s + \ldots + s^q}, \quad p \leq q \qquad (6/1)$$

are to be determined in such a way that for the given points of time $t_i$ the measured values of the step response are fitted as well as possible in the sense of a least sum of squares by the values of the step response calculated from the transfer function $\hat{G}(s)$. For starting the proposed method for the solution of the resulting nonlinear least squares problem it is necessary to know starting values for the parameters $b_0, \ldots, b_p, a_0, \ldots, a_{q-1}$. Assume that a subset of the time constants of the dynamic system is approximately known - a situation which is often present. Utilizing the relation (6/1) for the transfer function $\hat{G}(s)$ these time constants do not directly take effect on the parameters $b_i$ and $a_j$, but only in a complicated way of sums and products. This disadvantage vanishes if we write the transfer function from eq. (6/1) in another way

$$\hat{G}(s) = \sum_{j=1}^{q} \frac{A_j}{s - \lambda_j} . \qquad (6/2)$$

In the main the model (6/2) is mathematically equivalent to the model (6/1). But in the relation (6/2) there is a direct association between the approximately known time constants - our assumption - and the parameters of the model (6/2). Moreover the model (6/2) has the advantage that the step response and its derivatives with respect to the parameters can easily be calculated, namely by the aid of an analytical formula. But this is not true for the model (6/1).

We recognize that it is possible under certain circumstances to reduce the problem in the choice of starting points by an appropriate choice of the function $\underline{f}(\underline{x})$. Appropriate choice means here that we utilize a mathematically equivalent - or nearly equivalent - formulation of the same functional relation in the limits of the possible formulations.

Remembering the linear least squares problem where the problem of the choice of starting points plays no part it is advantageous to describe the functional relation by such a function $\underline{f}(\underline{x})$ which linearly depends on as many parameters as possible. As recognized by the above considered example this is eventually attainable by utilizing an appropriate mathematical formulation for the function $\underline{f}(\underline{x})$. The advantage of a formulation for which the function $\underline{f}(\underline{x})$ linearly depends on as many parameters as possible is derived from the fact that the proposed method is always stable independent on the number of parameters to be modified. That means that it is also stable if we only modify a subset of all parameters. In the case that some parameters appear linearly in the function $\underline{f}(\underline{x})$ we only need starting points for the parameters on which the function $\underline{f}(\underline{x})$ depends nonlinearly because the starting points for the other parameters can be calculated by the solution of a linear least squares problem if we consider the parameters on which the function $\underline{f}(\underline{x})$ depends nonlinearly to be constants. The solution of this linear least squares problem can be determined in one step for arbitrary starting points. This solution is taken as starting point for the parameters on which the function $\underline{f}(\underline{x})$ depends linearly when solving the complete nonlinear problem. For the solution of this complete nonlinear least squares problem we utilize the given and the calculated values as the components of the starting point $\underline{x}^0$. In this way we save the specification of several starting values which are determined by the easy task of solving a linear least squares problem. The separation of the nonlinear least squares problem in a linear and a nonlinear one can also be applied during the processing of the proposed iterative method, that means in any stage of the iteration. This strategy can save a considerable amount of computation time (see section 7.3).

Assume we find a point $\underline{x}^*$ for which the gradient of the sum of squares vanishes. Then we have to decide whether the sum of squares really becomes minimal at this point. This can be examined - if necessary - by considering the matrix $\underline{H}(\underline{x}^*)$ from eq. (3.1/3). For a minimum it is sufficient that the matrix $\underline{H}(\underline{x}^*)$ is positive definite. This property can numerically be examined for instance with the help of the Cholesky decomposition, if we assume that the matrix $\underline{H}$ is symmetric which is usually true. If we find out that the sum of squares becomes minimal for $\underline{x}^*$ so we are interested in knowing whether we have reached a global or a local minimum. If we have some supplementary informations about the parameters $\underline{x}^*$ which we expect as solution this question can often be answered. Supposing that these supplementary informations are not available the usual proceeding is to restart the numerical method for the solution of nonlinear least squares problems with rather different starting points $\underline{x}^0$. If we always reach the same value $\underline{x}^*$ we can be sure to a certain extent that we have found the global minimum. If we find several different points $\underline{x}^*$ we choose that point for which the sum of squares is minimal and define it to be the point $\underline{x}^*$ we are looking for.

# 7. APPLICATIONS OF THE PROPOSED METHOD FOR THE SOLUTION OF NONLINEAR LEAST SQUARES PROBLEMS

In this chapter we want to show how several versions of the proposed method for the solution of nonlinear least squares problems can be applied in order to solve problems from various fields which yield nonlinear least squares problems when formulated in an appropriate manner. In section 7.1 we give a short description of the problems which are considered in this chapter. The functions $f_i(\underline{x})$ and their derivatives with respect to the parameters $x_j$ necessary for the solution of the resulting nonlinear least squares problems are derived in section 7.2. Finally in section 7.3 we give numerical examples for each problem treated in this chapter. These numerical examples shall demonstrate how a problem must be prepared in order to apply one of the presented numerical methods for its solution. Moreover the performance of several versions of the proposed method is studied among other things in comparison with Hartley's method.

## 7.1 Formulation of the problems

Often we want to study the performance of a given transfer system. If it is not possible to perform these studies with the real system - for instance because the reliability in service or the plant conditions do not allow it - we are depending on a simulation of the system. Usually this simulation is performed on a digital or an analog computer. In order to simulate the system we need a mathematical model of the transfer system under consideration that means we need a mathematical description which reproduces the performance of the real system as well as possible. The mathematical models may be very different. For instance it comes into question a system of linear or nonlinear differential equations, one linear or nonlinear differential equation of higher order, a system of difference equations, and so on. For the interests of control theory we often look for a description of the transfer system in the form of a transfer function $G(s)$. Indifferent which kind of mathematical model is used we must enter into the question how to determine such a mathematical model. There are two principal possibilities. The first possibility consists in the so-called theoretical analysis [1]. When applying it we derive a mathematical model of the dynamic system by the aid of the physical laws - for instance the conservation laws of mass, energy, impulse, charge, and so on.

Often the dynamic systems for which we want to derive a mathematical model are so complex that the theoretical analysis is not practicable. Then we must turn over to the second possibility, the so-called experimental analysis. When applying this method we try to derive a mathematical model of the dynamic system from measured values of the input and output of this system. In order to perform the experimental analysis we usually need numerical methods. There is a lot of well-known methods. A detailed representation may be found for instance in [2, 3]. The problem of the experimental analysis can be formulated as a nonlinear least squares problem. In that case the measured values of the output are the given fixed values $y_i$ gathered up in the vector $\underline{y}$. These values are to be fitted by the values of a mathematical model whose structure is given but which still depends on some parameters. Such a mathematical model may be described by functions $f_i(\underline{x})$. One possibility for the solution of this problem consists in the determination of the parameter vector $\underline{x}$ so that the sum of squares becomes minimal. The measured values may be values in the time domain - for instance those of a step response or an impulse response - or values in the frequency domain - for instance those from a Bode plot, that means values of a magnitude or phase angle plot. Appropriate functions $f_i(\underline{x})$ for the solution of these problems are derived in section 7.2. From these functions $f_i(\underline{x})$ we can obtain a description of the transfer system in the form of a transfer function G(s) we are often interested in.

Another problem which arises very often in practical applications is the design of transfer systems belonging to a given transfer system so that the resulting transfer system shows a desired performance. In order to make clear what we mean we consider the following situation.

Given a certain transfer system I with input u and output f as shown in figure 7/1.

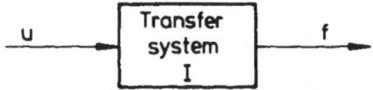

Fig. 7/1: Transfer system I with input u and output f.

If we utilize a certain input u we obtain a certain output f. Often it happens that the output f of the originally given transfer system I does not show the performance which we desire for a given input u. In order to overcome this drawback we can modify the given transfer system in such a way that we plan a dynamic transfer system II described for instance by a transfer function H(s). This transfer system II is connected with the given transfer system I in the way shown in figure 7/2. This structure is usually called open-loop control and is sometimes sufficient to achieve the desired performance of the system.

Fig. 7/2: Open-loop control.

If we assume that the structure of the transfer system II - described by its transfer function H(s) - is given but that the performance of this transfer system II still depends on several not yet determined parameters $x_j$ we have the chance to fix these parameters in such a way that the output f of the resulting transfer system (transfer system I + transfer system II) from figure 7/2 fits the desired output y as well as possible. If we consider only certain discrete values $y_i$ and $f_i$ of the desired output y and of the real output f we obtain a nonlinear least squares problem which can be solved by the methods described in preceding chapters. These values $y_i$ of the desired output y may be values for instance in the time domain or the frequency domain.

Often the configuration from figure 7/2 is not sufficient to reach the goal that the output f fits the desired output y well, especially if we pay attention to the fact that we usually have disturbances which attack the resulting transfer system and that the individual components of the transfer system can vary more or less. Then it is necessary to turn over to a feedback control as shown in figure 7/3.

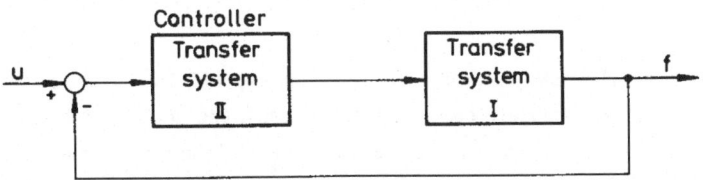

Fig. 7/3: Feedback control.

If the transfer system II is used in such a feedback control as shown in figure 7/3 it is usually called a controller. Such a controller can often be described by a transfer function H(s).

One of the main problems in control theory consists in designing an appropriate controller. Sometimes the problem is to design a controller in such a way that the output f of the feedback control shows a desired performance y when utilizing a certain input u. This requirement may be formulated by the aid of specifications in the time or frequency domain. If we assume that the structure of the controller is given but that its transfer function H(s) for instance still depends on some parameters $x_j$ and if we demand that certain values $f_i$ of the output f fit the given values $y_i$ of the desired output y as well as possible in the sense of a least sum of squares we have to solve a nonlinear least squares problem. We see that it is possible to design a controller by solving a nonlinear least squares problem that means by optimizing the parameters of the controller. In order to do this we often need a mathematical model of the originally given transfer system. Such a mathematical model can be derived for instance by the above described method of the experimental analysis. So the design of an appropriate controller can demand the solution of two nonlinear least squares problems.

A further problem which can be solved with the help of a formulation as a nonlinear least squares problem is the problem of the so-called simplification of dynamic systems. When considering the problem of the simplification of dynamic systems we assume that we already know a mathematical model of the given dynamic system. Furthermore we assume that this mathematical model is too complex for further investigations. That means the difficulties when examining it analytically respectively the

numerical effort when simulating it on a digital or analog computer are so immense
that we are looking for a simplified model which reproduces the main properties of
the originally given mathematical model. In this connexion we must clear up the
question under which conditions we want to speak of a simplified mathematical model.
This question cannot be answered in a general valid manner. It depends on the actual
problem under consideration. If the mathematical model of the originally given
transfer system is a transfer function G(s) whereby the degree of the polynomial of
the denominator is n then we speak of a simplified mathematical model if the degree
of the polynomial of the denominator of the transfer function of the simplified model
is less than n. In this sense we want to use the notion simplification of dynamic
systems here.

There is a lot of well-known methods to solve this problem. A survey is given for instance in [4, 5]. This problem can also be formulated as a nonlinear least squares
problem. For that purpose we compute values $y_i$ belonging to the originally given
mathematical model. These values may be values for instance of a step response, an
impulse response, a magnitude or phase angle plot, and so on. These computed values
$y_i$ are considered as the given fixed values $y_i$ which have to be fitted by the values
$f_i$ of a simplified mathematical model whereby this mathematical model still depends
on some not yet determined parameters $x_j$. These parameters are determined in such a
way that the given fixed values $y_i$ are as well as possible fitted by the values $f_i$.
By the aid of this formulation the problem of the simplification of dynamic systems
can be transformed to a nonlinear least squares problem. We see that the problem of
the simplification of dynamic systems is related to the problem of the experimental
analysis because it is unimportant for a numerical method whether the given fixed
values $y_i$ are measured or calculated from a given mathematical model.

Sometimes we have to solve the following problem which arises in optimal control
theory. Given a certain transfer system described by a mathematical model, for
instance in the form of an open-loop transfer function G(s). We look for such a
control function u(t) so that the output of the transfer system shows a desired performance. This problem can be solved for instance in such a way that we prescribe the
structure of the control function still depending on some parameters. Such a control
function u(t) may be a function which can only accept the two values $u_1$ and $u_2$ while
the switching instants between these two values are not fixed. Then these switching
instants are the parameters $x_j$ of the control function u(t). When utilizing such a
function u(t) as input of a transfer system we obtain a certain output f(t) for a
given set of parameters $x_j$. Choosing other values for the parameters $x_j$ we shall in
general obtain another output f(t) of the transfer system. Therefore the output f(t)
depends on the parameters $x_j$ of the function u(t) used as input. Now we assume that a
prescribed performance of the output y(t) of the transfer system is given. We only
want to consider certain discrete values $y_i$ of it. Then we can look for such a set of

parameters $x_j$ of the function $u(t)$ used as input of the transfer system so that the given fixed values $y_i$ of the prescribed output $y$ are as well as possible fitted by the values $f_i$ of the real output of the transfer system. The problem in this formulation also yields a nonlinear least squares problem.

The problems formulated in this section are to be solved examplarily. As they are formulated as nonlinear least squares problems we can try to solve them by the aid of the methods presented in the preceding chapters. Before we can apply these methods it is necessary to specify the functions $f_i(\underline{x})$ and to give relations for their derivatives with respect to the parameters $x_j$.

## 7.2    Specification of the functions $f_i(\underline{x})$

As pointed out in chapter 6 it is convenient to utilize "appropriate" functions $f_i(\underline{x})$, for instance in order to obtain a "good" starting point $\underline{x}^0$ from supplementary informations. Furthermore we are interested in a simple and quick calculation of the values of the functions $f_i(\underline{x})$ and of their derivatives with respect to the parameters $x_j$ as given by eq. (3.1/12).

In particular it is important to specify the derivatives analytically because their determination with the help of a numerical approximation is very expensive and in general pretty inaccurate. But just an exact determination of the derivatives within the accuracy of a digital computer is required because the sequence of the selection of the vectors $\underline{d}_j^k$ for the building up of the matrix $\underline{D}^k$ and with it the determination of the direction vector $\underline{r}^k$ depends on these vectors $\underline{d}_j^k$, that means it depends on the derivatives of the function $\underline{f}(\underline{x})$ with respect to the parameters $x_j$. As the proposed new method is essentially based on a suitable selection of the vectors $\underline{d}_j^k$ for the building up of the matrix $\underline{D}^k$ the formulas with which we calculate the vectors $\underline{d}_j^k$ have to be given analytically in order to reduce the errors to a minimum when determining the vectors $\underline{d}_j^k$.

In the following subsections we shall show up how these requirements can be satisfied for the problems formulated in section 7.1.

## 7.2.1 Functions $f_i(\underline{x})$ for the fit of a magnitude and phase angle plot

Now we consider the problem that the given fixed values $y_i$, $i = 1, \ldots, m$ are the values of a measured or calculated magnitude and phase angle plot given for certain frequencies $\omega_i$, $i = 1, \ldots, m$. In general these values $y_i$ can consist of values of a magnitude plot as well as of a phase angle plot. It is not necessary that for a certain frequency $\omega_i$ the value of the magnitude plot and the value of the phase angle plot are both given although this case will occur very often. It is also permitted that for instance only values of the magnitude plot are given. This problem occurs in technical applications if only the magnitude plot of a transfer function is of interest while the phase angle plot plays no part. Such problems are also covered by the functions $f_i(\underline{x})$ derived in the sequel.

Now we want to consider the problem that the given fixed values $y_i$ of a magnitude and phase angle plot are to be fitted - in the sense of a least sum of squares - by the values of a magnitude and phase angle plot of a linear time-invariant mathematical model. Such a mathematical model can be described by a transfer function $\hat{G}(s)$. For this transfer function $\hat{G}(s)$ we choose the following mathematical formulation

$$\hat{G}(s) = \frac{V}{s^\rho} \frac{\displaystyle\prod_{j=1}^{\tilde{n}_r} (1+\frac{s}{\tilde{\omega}_j}) \prod_{j=1}^{\tilde{n}_k} (1+\frac{2\tilde{\zeta}_j s}{\tilde{\omega}_{nj}} + \frac{s^2}{\tilde{\omega}_{nj}^2})}{\displaystyle\prod_{j=1}^{n_r} (1+\frac{s}{\omega_j'}) \prod_{j=1}^{n_k} (1+\frac{2\zeta_j' s}{\omega_{nj}'} + \frac{s^2}{\omega_{nj}'^2})} \qquad (7.2.1/1)$$

which is sometimes called Bode form [6]. In eq. (7.2.1/1) the numerator and the denominator of the transfer function $\hat{G}(s)$ are written as the product of linear factors, of quadratic factors - in which the contributions of conjugate complex roots are gathered up -, and of a factor $s^\rho$ which contains the contributions produced by the roots $s = 0$. V is the so-called Bode gain.

The values of the magnitude plot belonging to the transfer function $\hat{G}(s)$ from eq. (7.2.1/1) are given by

$$|\hat{G}(j\omega)|_{dB} = |V|_{dB} - |(j\omega)^\rho|_{dB} + \sum_{j=1}^{\tilde{n}_r} |1 + \frac{j\omega}{\tilde{\omega}_j}|_{dB} +$$

$$+ \sum_{j=1}^{\tilde{n}_k} |1 + 2\tilde{\zeta}_j \frac{j\omega}{\tilde{\omega}_{nj}} + \frac{(j\omega)^2}{\tilde{\omega}_{nj}^2}|_{dB} -$$

$$- \sum_{j=1}^{n_r} |1 + \frac{j\omega}{\omega_j'}|_{dB} - \sum_{j=1}^{n_k} |1 + 2\zeta_j' \frac{j\omega}{\omega_{nj}'} + \frac{(j\omega)^2}{\omega_{nj}'^2}|_{dB} \qquad (7.2.1/2)$$

and the values of the phase angle plot are given by

$$arc\{\hat{G}(j\omega)\} = arc\{V\} - arc\{(j\omega)^\rho\} + \sum_{j=1}^{\tilde{n}_r} arc\{1 + \frac{j\omega}{\tilde{\omega}_j}\} +$$

$$+ \sum_{j=1}^{\tilde{n}_k} arc\{1 + 2\tilde{\zeta}_j \frac{j\omega}{\tilde{\omega}_{nj}} + \frac{(j\omega)^2}{\tilde{\omega}_{nj}^2}\} - \sum_{j=1}^{n_r} arc\{1 + \frac{j\omega}{\omega_j'}\} -$$

$$- \sum_{j=1}^{n_k} arc\{1 + 2\zeta_j' \frac{j\omega}{\omega_{nj}'} + \frac{(j\omega)^2}{\omega_{nj}'^2}\} . \qquad (7.2.1/3)$$

If we utilize the known values $\omega_i$ for $\omega$ in the eqs. (7.2.1/2) and (7.2.1/3) we obtain the functions $f_i$ we are looking for. Additionally it is necessary to fix the values $\rho$, $\tilde{n}_r$, $\tilde{n}_k$, $n_r$, and $n_k$. Then the functions $f_i$ depend on the parameters

$$\underline{x} = (\tilde{\zeta}_1, \ldots, \tilde{\zeta}_{\tilde{n}_k}, \tilde{\omega}_{n1}, \ldots, \tilde{\omega}_{nn_k}, \zeta'_1, \ldots, \zeta'_{n_k},$$

$$\omega'_{n1}, \ldots, \omega'_{nn_k}, \tilde{\omega}_1, \ldots, \tilde{\omega}_{\tilde{n}_r}, \omega'_1, \ldots, \omega'_{n_r}, V)^T \quad (7.2.1/4)$$

whereby the number n of parameters to be determined is given by $\tilde{n}_r + 2\,\tilde{n}_k + n_r + 2\,n_k$ + 1. Because of the relation (7.2.1/1) we see that the components $x_j$ of the parameter vector $\underline{x}$ are all real-valued. The parameter vector $\underline{x}$ has to be determined with the proposed or another method in such a way that the given fixed values $y_i$ of the magnitude and phase angle plot are fitted as well as possible - in the sense of a least sum of squares $Q(\underline{x})$ - by the functions $f_i(\underline{x})$. As the functions $f_i$ depend nonlinearly on the parameters $x_j$ we have to solve a nonlinear least squares problem. In order to apply a numerical method for its solution it is necessary to calculate the values of the functions $f_i$ for a specific parameter vector $\underline{x}$ and the derivatives of these functions $f_i(\underline{x})$ with respect to the parameters $x_j$, $j = 1, \ldots, n$.

For the determination of the value of a function $f_i$ it is necessary to evaluate either eq. (7.2.1/2) or eq. (7.2.1/3) which depends on the fact whether the given value $y_i$ is a value of the magnitude or phase angle plot. This calculation makes no difficulties because the terms to be evaluated are not complicated. When considering the determination of the derivatives of the functions $f_i(\underline{x})$ with respect to the parameters $x_j$ the Bode form (7.2.1/1) of the transfer function is very advantageous. Because of the relations (7.2.1/2) and (7.2.1/3) we obtain formulas for the derivatives which can easily be evaluated. The derivatives depend on maximal two unknown parameters.

Before we derive the formulas for the derivatives of the functions $f_i(\underline{x})$ with respect to the parameters $x_j$ we want to introduce scaled frequencies instead of the originally given frequencies $\omega_i$. The formers are given by

$$\Omega_i = \frac{\omega_i}{\omega_{ref}} \quad . \quad (7.2.1/5)$$

In eq. (7.2.1/5) $\omega_{ref}$ means an arbitrary reference frequency. Instead of the parameters $\tilde{\omega}_1, \ldots, \tilde{\omega}_{\tilde{n}_r}, \tilde{\omega}_{n1}, \ldots, \tilde{\omega}_{nn_k}, \omega'_1, \ldots, \omega'_{n_r}, \omega'_{n1}, \ldots, \omega'_{nn_k}$ from eq. (7.2.1/1) we introduce the new parameters

$$\tilde{\Omega}_j = \frac{\omega_{ref}}{\tilde{\omega}_j} \quad (7.2.1/6)$$

$$\tilde{\Omega}_{nj} = \frac{\omega_{ref}}{\tilde{\omega}_{nj}} \quad (7.2.1/7)$$

$$\Omega'_j = \frac{\omega_{ref}}{\omega'_j} \quad (7.2.1/8)$$

$$\Omega'_{nj} = \frac{\omega_{ref}}{\omega'_{nj}} \quad . \quad (7.2.1/9)$$

By the choice of an appropriate value for $\omega_{ref}$ we can achieve that these new parameters - which have no physical unit of measure - are well suited for the processing on a digital computer. When considering the problem of the fit of values of a magnitude plot and when utilizing the above introduced new parameters we obtain the following relations for the derivatives of the functions $f_i$ with respect to these parameters

$$\frac{\partial f_i}{\partial \tilde{\Omega}_j} = 20 \log(e) \frac{\Omega_i^2 \tilde{\Omega}_j}{1+(\Omega_i \tilde{\Omega}_j)^2} \quad , \quad j = 1, \ldots, \tilde{n}_r \quad , \tag{7.2.1/10}$$

$$\frac{\partial f_i}{\partial \tilde{\Omega}_{nj}} = 40 \log(e) \frac{\Omega_i^2 \tilde{\Omega}_{nj}[(2\tilde{\zeta}_j)^2 - 1 + (\Omega_i \tilde{\Omega}_{nj})^2]}{[1-(\Omega_i \tilde{\Omega}_{nj})^2]^2 + (2\tilde{\zeta}_j \Omega_i \tilde{\Omega}_{nj})^2} \quad , \quad j=1, \ldots, \tilde{n}_k, \tag{7.2.1/11}$$

$$\frac{\partial f_i}{\partial \tilde{\zeta}_j} = 80 \log(e) \frac{\tilde{\zeta}_j (\Omega_i \tilde{\Omega}_{nj})^2}{[1-(\Omega_i \tilde{\Omega}_{nj})^2]^2 + (2\tilde{\zeta}_j \Omega_i \tilde{\Omega}_{nj})^2} \quad , \quad j=1, \ldots, \tilde{n}_k, \tag{7.2.1/12}$$

$$\frac{\partial f_i}{\partial \Omega_j'} = -20 \log(e) \frac{\Omega_i^2 \Omega_j'}{1+(\Omega_i \Omega_j')^2} \quad , \quad j = 1, \ldots, n_r \quad , \tag{7.2.1/13}$$

$$\frac{\partial f_i}{\partial \Omega_{nj}'} = -40 \log(e) \frac{\Omega_i^2 \Omega_{nj}'[(2\zeta_j')^2 - 1 + (\Omega_i \Omega_{nj}')^2]}{[1-(\Omega_i \Omega_{nj}')^2]^2 + (2\zeta_j' \Omega_i \Omega_{nj}')^2} \quad , \quad j=1, \ldots, n_k, \tag{7.2.1/14}$$

$$\frac{\partial f_i}{\partial \zeta_j'} = -80 \log(e) \frac{\zeta_j' (\Omega_i \Omega_{nj}')^2}{[1-(\Omega_i \Omega_{nj}')^2]^2 + (2\zeta_j' \Omega_i \Omega_{nj}')^2} \quad , \quad j = 1, \ldots, n_k, \tag{7.2.1/15}$$

$$\frac{\partial f_i}{\partial V} = 20 \log(e) \frac{1}{V} \text{ for } V \neq 0 . \tag{7.2.1/16}$$

For the derivatives of the functions $f_i$ in the case of the fit of values of a phase angle plot we obtain

$$\frac{\partial f_i}{\partial \tilde{\Omega}_j} = \frac{\Omega_i}{1 + (\Omega_i \tilde{\Omega}_j)^2} \quad , \quad j = 1, \ldots, \tilde{n}_r, \tag{7.2.1/17}$$

$$\frac{\partial f_i}{\partial \tilde{\Omega}_{nj}} = \frac{2\tilde{\zeta}_j \Omega_i [1+(\Omega_i \tilde{\Omega}_{nj})^2]}{[1-(\Omega_i \tilde{\Omega}_{nj})^2]^2 + (2\tilde{\zeta}_j \Omega_i \tilde{\Omega}_{nj})^2} \quad , \; j=1, \ldots, \tilde{n}_k, \quad (7.2.1/18)$$

$$\frac{\partial f_i}{\partial \tilde{\zeta}_j} = \frac{2\Omega_i \tilde{\Omega}_{nj} [1-(\Omega_i \tilde{\Omega}_{nj})^2]}{[1-(\Omega_i \tilde{\Omega}_{nj})^2]^2 + (2\tilde{\zeta}_j \Omega_i \tilde{\Omega}_{nj})^2} \quad , \; j = 1, \ldots, \tilde{n}_k, \quad (7.2.1/19)$$

$$\frac{\partial f_i}{\partial \Omega'_j} = - \frac{\Omega_i}{1 + (\Omega_i \Omega'_j)^2} \quad , \; j = 1, \ldots, n_r, \quad (7.2.1/20)$$

$$\frac{\partial f_i}{\partial \Omega'_{nj}} = - \frac{2\zeta'_j \Omega_i [1+(\Omega_i \Omega'_{nj})^2]}{[1-(\Omega_i \Omega'_{nj})^2]^2 + (2\zeta'_j \Omega_i \Omega'_{nj})^2} \quad , \; j = 1, \ldots, n_k, \quad (7.2.1/21)$$

$$\frac{\partial f_i}{\partial \zeta'_j} = - \frac{2\Omega_i \Omega'_{nj} [1-(\Omega_i \Omega'_{nj})^2]}{[1-(\Omega_i \Omega'_{nj})^2]^2 + (2\zeta'_j \Omega_i \Omega'_{nj})^2} \quad , \; j = 1, \ldots, n_k, \quad (7.2.1/22)$$

$$\frac{\partial f_i}{\partial V} = 0 \quad \text{for } V \neq 0 . \quad (7.2.1/23)$$

We see that the relations which have to be evaluated in the eqs. (7.2.1/10 - 15) and (7.2.1/17 - 22) emerge several times so that the calculations of the derivatives are not very expensive. Moreover some of these terms can be utilized for the determination of the value of a function $f_i$. Therefore it is convenient to calculate the values of the functions $f_i$ and the corresponding derivatives of the functions $f_i$ with respect to the parameters $x_j$ in one subroutine on a digital computer.

As we recognize it is possible to give analytical formulas for the derivatives of the functions $f_i$. Therefore we are able to calculate the vectors $\underline{d}_j^k$ exactly within the limits of the accuracy of the utilized digital computer.

For the application of the proposed method it is necessary to input the values for $\rho$, $\tilde{n}_r$, $\tilde{n}_k$, $n_r$, and $n_k$ which appear in the Bode form of the transfer function $\hat{G}(s)$ we are interested in. Moreover we need a starting point $\underline{x}^0$. Because of the chosen formulation of the transfer function $\hat{G}(s)$ such a starting point $\underline{x}^0$ can often be derived from supplementary informations or simply by considering the given fixed values $y_i$ in the Bode plot. Because of the introduced scaled parameters from eqs. (7.2.1/6 - 9) it is only necessary to choose a part of these parameters as multiples of the reference frequency $\omega_{ref}$. In principle, it is possible to choose any arbitrary frequency as $\omega_{ref}$. However it is convenient to select this frequency out of the given range of frequencies $[\omega_{min}, \omega_{max}]$ to which the given frequencies $\omega_i$ belong.

When determining the vector $\underline{x}^*$ by the aid of one of the above-mentioned numerical methods there can occur problems concerning the determination of the Bode gain V from eq. (7.2.1/1) because the derivatives (7.2.1/16) and (7.2.1/23) do not exist for V = 0. This problem can often be avoided by determining the Bode gain V in another way. Then it is no longer a parameter to be determined. If we consider the problem of the simplification of dynamic systems we are often interested that the Bode gains of the given and the simplified mathematical model are identical. In these cases the Bode gain can be considered as a known value. When considering the problem of the experimental analysis we often have informations about the Bode gain in the range of low frequencies, sometimes even for $\omega = 0$. From this we can determine the value of the Bode gain V or at least we can derive a good starting point for it. If we know for instance that the Bode gain is surely non-negative we can pay attention to it by introducing a new parameter $\tilde{V}$ instead of V according to

$$V = \tilde{V}^2. \qquad (7.2.1/24)$$

This new parameter $\tilde{V}$ together with the other parameters is determined by the aid of a numerical method for the solution of nonlinear least squares problems. After having performed this fit the Bode gain V from eq. (7.2.1/1) is calculated according to eq. (7.2.1/24).

In order to obtain the transfer function $\hat{G}(s)$ from eq. (7.2.1/1) it is necessary to calculate the parameters $\tilde{\omega}_j$, $\tilde{\omega}_{nj}$, $\omega'_j$, and $\omega'_{nj}$ from the parameters $\tilde{\Omega}_j$, $\tilde{\Omega}_{nj}$, $\Omega'_j$, and $\Omega'_{nj}$ with the help of the relations (7.2.1/6 - 9). Then we know all parameters which appear in the Bode form (7.2.1/1) of the transfer function $\hat{G}(s)$ so that we have solved the problem to determine a linear time-invariant mathematical model from the given fixed values $y_i$ of a magnitude and phase angle plot.

## 7.2.2  Functions $f_i(\underline{x})$ for the fit of a step response

The step response is the output $y(t)$ of a dynamic system if the input $u(t)$ is the step function $\sigma(t-T)$ and if all initial conditions of the output $y(t)$ are zero at $t = T$. The step function $\sigma(t-T)$ is defined by

$$\sigma(t-T) = \begin{cases} 1 \text{ for } t > T \\ 0 \text{ for } t \le T \end{cases} \qquad (7.2.2/1)$$

A graphical plot of the step function is shown in figure 7/4.

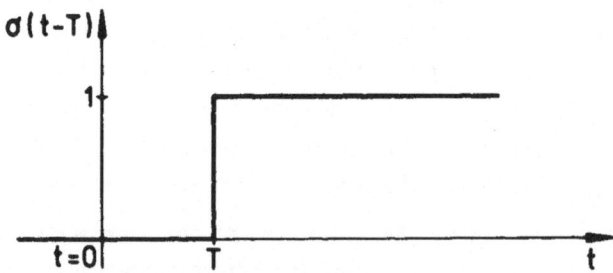

Fig. 7/4: Graphical plot of the step function.

If such a step function is the input u(t) of a dynamic system we obtain the step response y(t) as output. A typical step response is represented in figure 7/5.

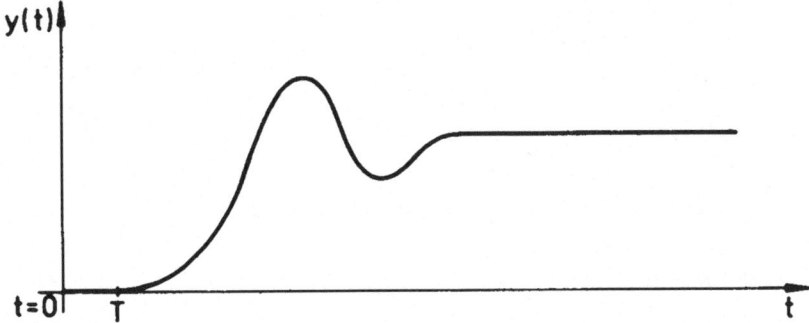

Fig. 7/5: Typical step response of a dynamic system.

Assume that this step response is measured or calculated for certain points of time $t_i$, i = 1, ..., m which need not be equidistant. So we obtain a finite number of values of the step response belonging to the selected points of time $t_i$. These selected values of the step response are the values $y_i$ which are to be fitted. For the step response from figure 7/5 a possible subset of selected values of the step response is represented in figure 7/6.

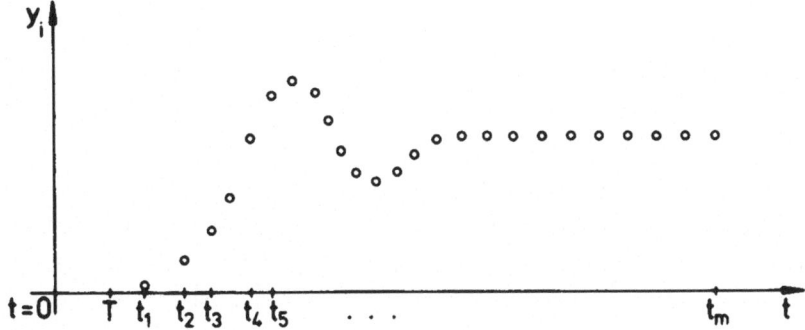

Fig. 7/6: Values of the step response from figure 7/5 for selected points of time $t_i$.

The given fixed values $y_i$ of the step response are to be fitted - in the sense of a least sum of squares Q - by the values of a step response belonging to a linear time-invariant transfer system. It is our goal to obtain a linear time-invariant mathematical model, for instance in the form of a transfer function $\hat{G}(s)$.

As we want to fit the given fixed values $y_i$ of the step response by the values of the step response of a time-invariant mathematical model it means no loss of generality if we set

$$T = 0 \qquad\qquad (7.2.2/2)$$

that means the input u(t) of the transfer system is the step function σ(t). In this case it is well-known [7] that the step response f(t) of a linear time-invariant mathematical model is given by

$$f(t) = A_0 + \sum_{j=1}^{n_r} A_j e^{-\tilde{\delta}_j t} \sigma(t) + \sum_{j=1}^{n_k} e^{-\delta_j t} (B_j \cos\omega_j t + C_j \sin\omega_j t)\sigma(t) \qquad (7.2.2/3)$$

if all eigenvalues are distinct and different from zero. In eq. (7.2.2/3) the value $n_r$ means the number of real eigenvalues while $n_k$ is the number of pairs of complex eigenvalues. As the step response of the dynamic system to be fitted is given only for certain points of time $t_i$ we only need the values of the step response from eq. (7.2.2/3) for these points of time $t_i$. Utilizing these values we obtain the functions $f_i$ given by

$$f(t_i) = f_i = A_0 + \sum_{j=1}^{n_r} A_j e^{-\tilde{\delta}_j t_i} + \sum_{j=1}^{n_k} e^{-\delta_j t_i} (B_j \cos\omega_j t_i +$$

$$+ C_j \sin\omega_j t_i), \quad i = 1, \ldots, m. \qquad (7.2.2/4)$$

If we gather up the parameters which appear in eq. (7.2.2/4) in the parameter vector $\underline{x}$ according to

$$\underline{x} = (C_1, \ldots, C_{n_k}, A_1, \ldots, A_{n_r}, A_0, B_1, \ldots, B_{n_k},$$

$$\omega_1, \ldots, \omega_{n_k}, \delta_1, \ldots, \delta_{n_k}, \tilde{\delta}_1, \ldots, \tilde{\delta}_{n_r})^T \qquad (7.2.2/5)$$

we see that the functions $f_i$ from eq. (7.2.2/4) depend nonlinearly on the components of this vector $\underline{x}$. Because of the relation (7.2.2/3) it is clear that all components of the parameter vector $\underline{x}$ are real-valued.

Now we want to solve the following problem: Determine the parameter vector $\underline{x}$ with the help of the proposed or another numerical method so that the given fixed values $y_i$ of the step response are fitted by the functions $f_i(\underline{x})$ as well as possible in the sense of a least sum of squares $Q(\underline{x})$. In order to apply the proposed or one of the other numerical methods presented in chapter 3 we do not only need the values of the functions $f_i$ belonging to a certain parameter vector $\underline{x}$ but also the derivatives of these functions $f_i(\underline{x})$ with respect to the components $x_j$, $j = 1, \ldots, n$ of the parameter vector. The number n of these components is given by

$$n = 2 n_r + 4 n_k + 1 \qquad (7.2.2/6)$$

which follows immediately from eq. (7.2.2/5).

The derivatives can be derived from eq. (7.2.2/4). We obtain

$$\frac{\partial f_i}{\partial A_0} = 1 \quad , \qquad (7.2.2/7)$$

$$\frac{\partial f_i}{\partial A_j} = e^{-\tilde{\delta}_j t_i}, \qquad\qquad j = 1, \ldots, n_r, \qquad (7.2.2/8)$$

$$\frac{\partial f_i}{\partial \tilde{\delta}_j} = -A_j t_i\, e^{-\tilde{\delta}_j t_i}, \qquad\qquad j = 1, \ldots, n_r, \qquad (7.2.2/9)$$

$$\frac{\partial f_i}{\partial \delta_j} = -t_i e^{-\delta_j t_i}(B_j \cos\omega_j t_i + C_j \sin\omega_j t_i), \qquad j = 1, \ldots, n_k, \qquad (7.2.2/10)$$

$$\frac{\partial f_i}{\partial B_j} = e^{-\delta_j t_i} \cos\omega_j t_i, \qquad\qquad j = 1, \ldots, n_k, \qquad (7.2.2/11)$$

$$\frac{\partial f_i}{\partial C_j} = e^{-\delta_j t_i} \sin\omega_j t_i, \qquad\qquad j = 1, \ldots, n_k, \qquad (7.2.2/12)$$

$$\frac{\partial f_i}{\partial \omega_j} = t_i\, e^{-\delta_j t_i}(-B_j \sin\omega_j t_i + C_j \cos\omega_j t_i), \qquad j = 1, \ldots, n_k. \qquad (7.2.2/13)$$

We recognize that the different terms which have to be determined appear in the eqs. (7.2.2/7 - 13) several times. Hence it follows that the calculation of the derivatives is not very expensive. Moreover it is possible to utilize many of the terms from eqs. (7.2.2/7 - 13) for the determination of the values of the functions $f_i$ so that it is convenient to plan one subroutine for the determination of the values of the functions $f_i$ and their derivatives when performing it on a digital computer. Furthermore we recognize by the eqs. (7.2.2/7 - 13) that we have analytical formulas for the derivatives of the functions $f_i(\underline{x})$ with respect to $x_j$. That means, it is possible to determine the vectors $\underline{d}_j^k$ exactly within the limits of the accuracy of the utilized digital computer.

Before applying a numerical method we need values for $n_r$ and $n_k$. These depend on the mathematical model which we want to obtain. Moreover we need a starting point $\underline{x}^0$. For the parameters on which the functions $f_i(\underline{x})$ from eq. (7.2.2/4) depend linearly we need no starting values if we make use of the method described in chapter 6 for the determination of these parameters by the solution of a linear least squares problem. In this case we only need starting values for those parameters on which the functions $f_i(\underline{x})$ depend nonlinearly. Such starting values can often be obtained by supplementary informations - for instance the knowledge of approximate values of time constants - or simply by considering the given fixed values $y_i$ of the step response. From these considerations we can obtain approximate values for the time constants $1/\tilde{\delta}_j$, the damping coefficients $\delta_j$, and the damped natural frequencies $\omega_j$. Therefore we can state that the functions $f_i$ given by eq. (7.2.2/4) are "well appropriate" for the solution of the problem under consideration.

Now we have performed all preparations for the application of a numerical method for the solution of nonlinear least squares problems concerning the above described problem of the fit of the values $y_i$ of a step response.

Assume we have applied a numerical method for the solution of the resulting nonlinear least squares problem. Then as a result of it we obtain a parameter vector $\underline{x}^*$ under the supposition that the utilized method converges. With this vector $\underline{x}^*$ we have found the best values for the parameters $x_j$ from eq. (7.2.2/5) in the sense of a least sum of squares. But these values are not our main goal, rather it is the corresponding transfer function $\hat{G}(s)$. This transfer function can easily be derived from the knowledge of the parameter vector $\underline{x}^*$.

In order to show this we must pay attention to the fact that the function $f(t)$ is the step response of a linear time-invariant mathematical model described by its transfer function $\hat{G}(s)$. Therefore the Laplace transform of $f(t)$ given by

$$\mathcal{L}\{f(t)\} = \frac{\hat{G}(s)}{s} \ . \tag{7.2.2/14}$$

Because of the supposed linearity of the transfer system we can assume in eq. (7.2.2/14) without loss of generality that the input $u(t)$ is the unit step $\sigma(t)$. From eq. (7.2.2/14) we obtain

$$\hat{G}(s) = s\,\mathcal{L}\{f(t)\} \ . \tag{7.2.2/15}$$

The Laplace transform $\mathcal{L}\{f(t)\}$ of the function $f(t)$ can easily be derived from eq. (7.2.2/3). With it we obtain from eq. (7.2.2/15)

$$\hat{G}(s) = A_0 + \sum_{j=1}^{n_r} \frac{A_j s}{s+\tilde{\delta}_j} + \sum_{j=1}^{n_k} \frac{B_j s^2 + (B_j\delta_j + C_j\omega_j)s}{(s+\delta_j)^2 + \omega_j^2} \ . \tag{7.2.2/16}$$

By summing up the partial fractions we obtain the transfer function $\hat{G}(s)$ in the form

$$\hat{G}(s) = \frac{b_0 + b_1 s + \ldots + b_{n_r+2n_k} s^{n_r+2n_k}}{a_0 + a_1 s + \ldots + s^{n_r+2n_k}} \ . \tag{7.2.2/17}$$

As we recognize by eq. (7.2.2/17) the coefficient $b_{n_r+2n_k}$ will be different from zero in general. But this means that the step response $f(t)$ is not continuous at the point $t = 0$. Often we know that the step response of the dynamic system from which the given fixed values $y_i$ result is continuous at $t = 0$. Sometimes we wish that the step response $f(t)$ also possesses this property. This can easily be achieved by modifying the relation (7.2.2/3) for the step response of the linear time-invariant mathematical model a little bit. For that purpose we calculate the value of the step response $f(t)$ at $t = 0$. It is given by

$$f(0) = A_0 + \sum_{j=1}^{n_r} A_j + \sum_{j=1}^{n_k} B_j \ . \tag{7.2.2/18}$$

If we demand that the step response is continuous at $t = 0$ it must have the same value as the given step response at t = 0. By an appropriate transformation of the values $y_i$ we can always achieve that this value is zero. Therefore we obtain the relation

$$A_0 + \sum_{j=1}^{n_r} A_j + \sum_{j=1}^{n_k} B_j = 0 \qquad (7.2.2/19)$$

from eq. (7.2.2/18). With the help of this relation we can eliminate one unknown parameter from eq. (7.2.2/19), for instance the parameter $A_0$. By proceeding in this way we save the determination of one parameter and it is guaranteed that the step response (7.2.2/3) is continuous at t = 0. The relations for the derivatives of the functions $f_i(\underline{x})$ with respect to the parameters $x_j$ as they are given by the eqs. (7.2.2/7 - 13) must slightly be modified but they are preserved on the whole so that the requirement that the step response f(t) is continuous at t = 0 means no complication of the problem.

Often we want to guarantee that the mathematical model obtained from the given fixed values $y_i$ of a step response is stable. That means the mathematical model shall possess the property that every bounded input produces a bounded output. As it is well-known a necessary and sufficient condition for this is that all values $\tilde{\delta}_j$ and $\delta_j$ are positive. By introducing new parameters $\tilde{\delta}'_j$ and $\delta'_j$ instead of the parameters $\tilde{\delta}_j$ and $\delta_j$ in the relation (7.2.2/4) for the functions $f_i(\underline{x})$, we can force that the values $\tilde{\delta}_j$ and $\delta_j$ are always nonnegative if we set

$$\tilde{\delta}_j = \tilde{\delta}'^2_j, \; j = 1, \ldots, n_r \qquad (7.2.2/20)$$

and

$$\delta_j = \delta'^2_j, \; j = 1, \ldots, n_k. \qquad (7.2.2/21)$$

So the stability of the mathematical model described by its transfer function $\hat{G}(s)$ is guaranteed on the supposition that all values $\tilde{\delta}'_j$ and $\delta'_j$ are different from zero. The introduction of these new parameters causes the relations for the derivatives of the functions $f_i(\underline{x})$ with respect to $x_j$ to change a little bit but the new relations are not more complicated than those from eqs. (7.2.2/7 - 13).

The problem of determining the parameters $x_j$ in such a way that the resulting transfer system is stable is a so-called nonlinear minimization problem with constraints referring to some parameters $x_j$. The constraints are inequalities of the type

$$\tilde{\delta}_j > 0 \text{ respectively } \delta_j > 0 \; . \qquad (7.2.2/22)$$

By the aid of the relations (7.2.2/20) and (7.2.2/21) this constrained nonlinear minimization problem is transformed to an unconstrained nonlinear minimization problem which can be solved by numerical methods derived for the solution of nonlinear least squares problems without constraints, for instance the proposed new method.

Until now we only consider mathematical models which yield step responses f(t) described by eq. (7.2.2/3). Sometimes it happens that the mathematical model must obtain an eigenvalue at s = 0. Then the step response f(t) from eq. (7.2.2/3) is unadequate. In this case we must add a term E·t in the step response from eq. (7.2.2/3). This supplementary term can easily be taken up in eq. (7.2.2/3). The analytical formula of the derivative belonging to this supplementary term with respect to the parameter E can be derived so that we can also treat this problem with the above described tools.

We have not yet considered problems in which the transfer function $\hat{G}(s)$ shall obtain multiple eigenvalues. Such problems are very interesting from a theoretical point of view but not from a practical point of view. The reason is that real dynamic systems never possess multiple eigenvalues in a strong mathematical sense. Therefore all practically relevant problems are covered by the above derived functions $f_i(\underline{x})$ and their derivatives with respect to the parameters $x_j$.

### 7.2.3 Functions $f_i(\underline{x})$ for the fit of an impulse response

The impulse response is the output y(t) of a dynamic system if the input u(t) is the impulse function δ(t) and all initial conditions are zero for t = 0. The impulse function δ(t) may be defined by

$$\delta(t) = \lim_{\substack{\Delta t \to 0 \\ \Delta t > 0}} [ \frac{\sigma(t) - \sigma(t-\Delta t)}{\Delta t} ] \qquad (7.2.3/1)$$

where σ(t) is the step function as defined by eq. (7.2.2/1). A graphical plot of the representation (7.2.3/1) for the impulse function for a finite value of Δt is shown in figure 7/7.

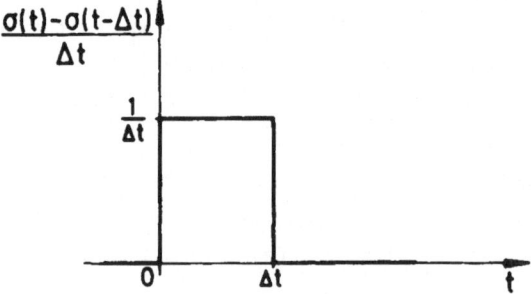

Fig. 7/7: One possible representation of the impulse function.

If the impulse function δ(t) is the input of a dynamic system we obtain the impulse response as output. A typical impulse response is shown in figure 7/8.

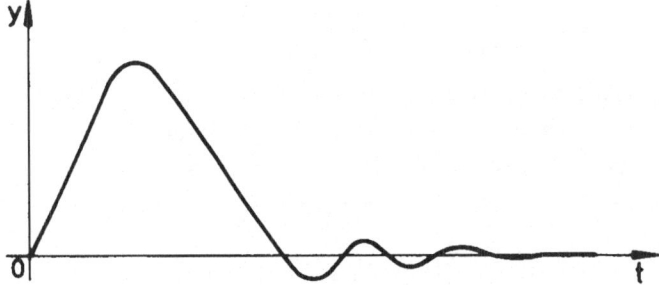

Fig. 7/8: Typical impulse response of a dynamic system.

On the analogy of the treatment of the step response in subsection 7.2.2 we assume that the values of this impulse response for certain points of time $t_i$, i = 1, ..., m are gathered up in the vector $\underline{y}$. Then we obtain certain values $y_i$ of the impulse response. In figure 7/9 such a set of values $y_i$ for the impulse response from figure 7/8 is shown.

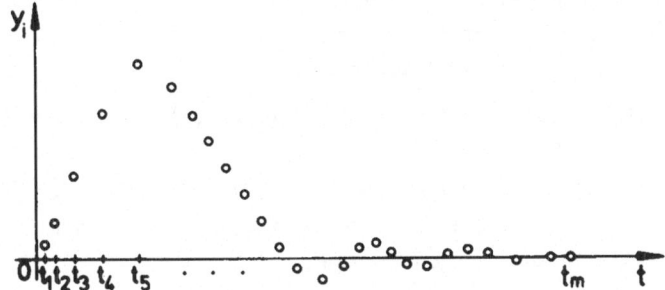

Fig. 7/9: Values $y_i$ of the impulse response from figure 7/8 for selected points of time.

The given fixed values $y_i$ of the impulse response are to be fitted as well as possible in the sense of a least sum of squares Q by values of an impulse response belonging to a linear time-invariant mathematical model. The impulse response of such a mathematical model is given by

$$f(t) = \sum_{j=1}^{n_r} A_j e^{-\tilde{\delta}_j t} + \sum_{j=1}^{n_k} e^{-\delta_j t} (B_j \cos\omega_j t + C_j \sin\omega_j t), \quad (7.2.3/2)$$

if all eigenvalues are distinct and different from zero [7]. In eq. (7.2.3/2) the value $n_r$ means the number of real eigenvalues while $n_k$ determines the number of pairs of complex eigenvalues. As described in subsection 7.2.2 we again introduce the parameter vector $\underline{x}$ according to eq. (7.2.2/5) - with the exception that $A_0$ does not emerge. Moreover we introduce the functions $f_i(\underline{x})$ and their derivatives with respect to the parameters $x_j$. The latter are given by the eqs. (7.2.2/8 - 13). We recognize that these functions are "well appropriate" for the solution of the resulting nonlinear least squares problem.

As result of the solution of this nonlinear least squares problem we obtain a parameter vector $\underline{x}^*$. Often we are interested in the transfer function $\hat{G}(s)$ belonging to the impulse response (7.2.3/2) with the determined parameter vector $\underline{x}^*$. As $f(t)$ is the impulse response its Laplace transform satisfies the relation

$$\mathcal{L}\{f(t)\} = \hat{G}(s). \tag{7.2.3/3}$$

Paying attention to eq. (7.2.3/2) we obtain for it

$$\mathcal{L}\{f(t)\} = \hat{G}(s) = \sum_{j=1}^{n_r} \frac{A_j}{s+\delta_j} + \sum_{j=1}^{n_k} \frac{B_j s + B_j \delta_j + C_j \omega_j}{(s+\delta_j)^2 + \omega_j^2} \quad . \tag{7.2.3/4}$$

By summing up the partial fractions in eq. (7.2.3/4) we obtain the transfer function $\hat{G}(s)$ in the form

$$\hat{G}(s) = \frac{b_0 + b_1 s + \ldots + b_{n_r+2n_k-1} s^{n_r+2n_k-1}}{a_0 + a_1 s + \ldots + s^{n_r+2n_k}} \quad . \tag{7.2.3/5}$$

If we demand that the impulse response belonging to $\hat{G}(s)$ is continuous at $t = 0$ and that the resulting transfer system described by the transfer function $\hat{G}(s)$ is stable the same considerations as in subsection 7.2.2 can be performed. Also the remarks from subsection 7.2.2 concerning an eigenvalue of $\hat{G}(s)$ at $s = 0$ and multiple eigenvalues of $\hat{G}(s)$ are valid in an analogous manner.

## 7.2.4 Functions $f_i(\underline{x})$ for the determination of a control function

In this subsection we want to consider the problem that the given fixed values $y_i$ are those of a prescribed output of an open-loop control system with input $u(t)$ and output $f(t)$ as represented in figure 7/10.

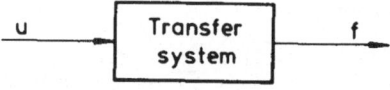

Fig. 7/10: Open-loop control (input u and output f).

As control function $u(t)$ we utilize only special functions with the two possible values $u = u_1$ and $u = u_2$. Such a function $u(t)$ is represented in figure 7/11.

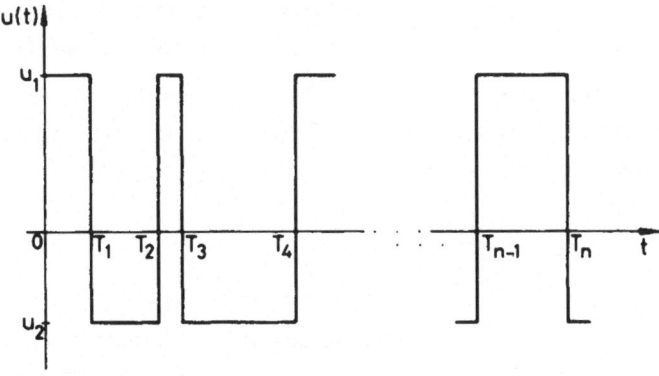

Fig. 7/11: Plot of a possible function u(t).

The switching instants in which the function u(t) changes its value from $u_1$ to $u_2$ or vice versa are denotated by $T_j$, j = 1, ..., n. We assume that the function u(t) has the value $u_1$ in the interval $[0, T_1)$ that means

$$u(t) = u_1 \text{ for } 0 \leq t < T_1 \qquad (7.2.4/1)$$

holds while u(t) vanishes for t < 0.

Such a control function u(t) is obtained for instance as output of a relay as represented in figure 7/12.

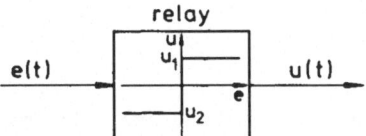

Fig. 7/12: Relay with input e and output u.

Such relay systems are of widespread usefulness. The relay as a power element for instance is simple to realize and strong. It is able to apply full power to a transfer system in a short time.

Corresponding to such an input u(t) we obtain a certain output f(t) of the transfer system under consideration. Naturally this output f(t) depends on the switching instants $T_j$. If we consider the output f(t) only at certain given points of time $t_i$, i = 1, ..., m we obtain m functions $f(t_i)$ which still depend on the parameters $T_j$. Now we want to solve the problem that the switching instants $T_j$ are determined in such a way that the given fixed values $y_i$ are fitted by the values of the functions $f_i$ as well as possible in the sense of a least sum of squares. The switching instants are to be determined by the solution of the resulting nonlinear least squares problem. In order to apply one of the above described numerical methods it is necessary to know the functions $f_i$ and their derivatives with respect to the parameters $T_j$.

At first we derive an analytical expression for the functions $f_i$. For that purpose we assume that the transfer system under consideration is described by an open-loop transfer function G(s) of the form

$$G(s) = \sum_{l=1}^{q} \frac{A_l}{s-\lambda_l} \quad . \tag{7.2.4/2}$$

This representation is valid if the eigenvalues of the mathematical model describing the transfer system are all distinct. In eq. (7.2.4/2) we admit real as well as complex eigenvalues $\lambda_l$. The representation (7.2.4/2) of the transfer function G(s) implies that the degree of the polynomial of the numerator is less than the degree of the polynomial of the denominator which is usually true when considering real transfer systems. If the transfer function G(s) is given in the form

$$G(s) = \frac{\sum_{k=1}^{p} b_k s^k}{\prod_{l=1}^{q} (s-\lambda_l)} = \frac{Z(s)}{\prod_{l=1}^{q} (s-\lambda_l)} \quad , \; p < q \tag{7.2.4/3}$$

it can be transformed to the representation (7.2.4/2) if we set

$$A_l = \frac{Z(\lambda_l)}{\prod_{\substack{k=1 \\ k \neq l}}^{q} (\lambda_l - \lambda_k)} \quad . \tag{7.2.4/4}$$

If we assume for the following considerations that all initial conditions of the output function f(t) are zero at t = 0 the Laplace transform of the output f(t) of the transfer system is given by

$$\mathcal{L}\{f(t)\} = G(s)\, \mathcal{L}\{u(t)\} \quad . \tag{7.2.4/5}$$

This relation is also valid if the function u(t) is not continuous and differentiable everywhere [8] - a situation which is present here. In order to evaluate eq. (7.2.4/5) we need an analytical formulation of the input u(t). For a function u(t) with n switching instants $T_j$ it is given by

$$u(t) = u_1[\sigma(t) - \sigma(t-T_1) + \sigma(t-T_2) - \ldots \pm \sigma(t-T_n)]$$

$$+ u_2[\sigma(t-T_1) - \sigma(t-T_2) + \ldots \mp \sigma(t-T_n)]$$

$$= u_1\sigma(t) + u_1 \sum_{j=1}^{n} (-1)^j \sigma(t-T_j) + u_2 \sum_{j=1}^{n} (-1)^{j+1} \sigma(t-T_j)$$

$$= u_1\sigma(t) + (u_1-u_2) \sum_{j=1}^{n} (-1)^j \sigma(t-T_j) \quad . \tag{7.2.4/6}$$

The Laplace transform of this function $u(t)$ is given by

$$\mathcal{L}\{u(t)\} = \frac{u_1}{s} + (u_1-u_2) \sum_{j=1}^{n} (-1)^j \frac{e^{-T_j s}}{s} \quad . \tag{7.2.4/7}$$

Utilizing the eqs. (7.2.4/2) and (7.2.4/7) in eq. (7.2.4/5) we obtain

$$\mathcal{L}\{f(t)\} = u_1 \sum_{l=1}^{q} \frac{A_l}{(s-\lambda_l)s} + (u_1-u_2) \left( \sum_{l=1}^{q} \frac{A_l}{s-\lambda_l} \right)$$

$$\left( \sum_{j=1}^{n} (-1)^j \frac{e^{-T_j s}}{s} \right) \quad . \tag{7.2.4/8}$$

From the last relation we can obtain the function $f(t)$ we are looking for. If we assume furthermore that the mathematical model has no eigenvalue at $s = 0$ the inverse transformation of eq. (7.2.4/8) can be performed with the help of a partial fraction expansion. We obtain

$$f(t) = u_1 \sum_{l=1}^{q} \frac{A_l}{\lambda_l} (e^{\lambda_l t}-1)\sigma(t) +$$

$$+ (u_1-u_2) \sum_{j=1}^{n} (-1)^j \left( \sum_{l=1}^{q} \frac{A_l}{\lambda_l} (e^{\lambda_l (t-T_j)}-1) \right) \sigma(t-T_j) \quad . \tag{7.2.4/9}$$

Introducing new coefficients according to

$$\frac{A_l}{\lambda_l} = r_l \tag{7.2.4/10}$$

and

$$\sum_{l=1}^{q} (-\frac{A_l}{\lambda_l}) = r_0 \tag{7.2.4/11}$$

we obtain from eq. (7.2.4/9)

$$f(t) = u_1 (r_0 + \sum_{l=1}^{q} r_l e^{\lambda_l t}) \sigma(t) +$$

$$+ (u_1-u_2) \sum_{j=1}^{n} (-1)^j (r_0 + \sum_{l=1}^{q} r_l e^{\lambda_l (t-T_j)}) \sigma(t-T_j). \tag{7.2.4/12}$$

If we utilize the given points of time $t_i$ in eq. (7.2.4/12) we obtain the functions $f_i$ we are looking for. These functions $f_i$ depend on the yet unknown parameters $T_j$, $j = 1, \ldots, n$. Now it is not advantageous to use the switching instants $T_j$ directly as parameters to be determined because these switching instants must satisfy the following inequalities

$$T_j \geq 0 \ , \ j = 1, \ldots, n \tag{7.2.4/13}$$

and

$$T_j \geq T_{j-1} \ , \quad j = 2, \ldots, n \tag{7.2.4/14}$$

because of the assumed structure of the function u(t) (compare figure 7/11). In order to satisfy these inequalities we introduce new parameters $x_j$, $j = 1, \ldots, n$ according to

$$T_j = \sum_{k=1}^{j} x_k^2 \ .$$ (7.2.4/15)

We see that the inequalities (7.2.4/13) and (7.2.4/14) are satisfied if we write the switching instants in the form of eq. (7.2.4/15). Utilizing the new parameters $x_j$ from eq. (7.2.4/15) we can write for the output f(t) of the transfer system under consideration

$$f(t) = u_1 \ (r_0 + \sum_{l=1}^{q} r_l e^{\lambda_1 t}) \sigma(t) +$$

$$+ (u_1 - u_2) \sum_{j=1}^{n} (-1)^j (r_0 + \sum_{l=1}^{q} r_l e^{\lambda_1 (t - \sum_{k=1}^{j} x_k^2)}) \sigma(t - \sum_{k=1}^{j} x_k^2). (7.2.4/16)$$

In order to apply one of the presented methods for the solution of the resulting non-linear least squares problem we need the derivatives of the functions $f_i$ with respect to the parameters $x_j$. From eq. (7.2.4/16) we can derive the following relation for these derivatives

$$\frac{\partial f_i}{\partial x_j} = 2(u_1 - u_2) x_j \sum_{r=j}^{n} (-1)^{r+1} \ [\ \sum_{l=1}^{q} r_l \lambda_1 e^{\lambda_1 (t_i - \sum_{k=1}^{r} x_k^2)}$$

$$\sigma(t_i - \sum_{k=1}^{r} x_k^2)], \ j = 1, \ldots, n.$$
(7.2.4/17)

With the eqs. (7.2.4/16) and (7.2.4/17) we have two analytical formulas at our disposal, namely for the determination of the value of the functions $f_i$ and their derivatives with respect to $x_j$.

Here we have a situation in which Meyer's method for the solution of the resulting nonlinear least squares problems may fail. From eq. (7.2.4/17) we see that a vector $\underline{d}_j$ which is given by eq. (3.1/11) vanishes if a switching instant $T_j$ is greater than the greatest value $t_i$ which appears. In this case Meyer's method fails because for its applicability it is necessary that all column vectors of the Jacobian matrix are different from the null vector (compare eq. (3.2/18)).

Often we have the situation that one of the two possible values of the function u(t) vanishes, for instance that

$$u_2 = 0$$ (7.2.4/18)

holds.

Then we obtain for the output of the transfer system from eq. (7.2.4/16)

$$f(t) = u_1(r_0 + \sum_{l=1}^{q} r_1 e^{\lambda_1 t})\sigma(t) +$$

$$+ u_1 \sum_{j=1}^{n} (-1)^j (r_0 + \sum_{l=1}^{q} r_1 e^{\lambda_1(t-\sum_{k=1}^{j} x_k^2)})\sigma(t-\sum_{k=1}^{j} x_k^2) \qquad (7.2.4/19)$$

and for its derivatives with respect to $x_j$

$$\frac{\partial f_i}{\partial x_j} = 2u_1 x_j \sum_{r=j}^{n} (-1)^{r+1}[\sum_{l=1}^{q} r_1 \lambda_1 e^{\lambda_1(t_i-\sum_{k=1}^{r} x_k^2)}\sigma(t_i-\sum_{k=1}^{r} x_k^2)] \quad . \qquad (7.2.4/20)$$

Until now, we have supposed that the initial conditions for the output function $f(t)$ are all zero at $t = 0-$, that means before applying the input function $u(t)$. Now we admit that these initial conditions are different from zero and that they are given by

$$\underline{f}(0-) = \begin{pmatrix} f(0-) \\ \dot{f}(0-) \\ \cdot \\ \cdot \\ \cdot \\ f^{(q-1)}(0-) \end{pmatrix} . \qquad (7.2.4/21)$$

It can be shown [8] that then the Laplace transform of the output $f(t)$ is given by

$$\mathcal{L}\{f(t)\} = G(s) \mathcal{L}\{u(t)\} + \frac{\underline{s}^T \underline{A} \underline{f}(0-)}{N(s)} \qquad (7.2.4/22)$$

where $N(s)$ is the denominator of the transfer function $G(s)$. The row vector $\underline{s}^T$ is given by

$$\underline{s}^T = (s^{q-1}, \ldots, s, 1) \qquad (7.2.4/23)$$

and the matrix $\underline{A}$ is given by

$$\underline{A} = \begin{pmatrix} 1 & & \\ a_{q-1} & 1 & \\ \cdots\cdots\cdots\cdots & & \\ a_1 & a_2 \cdots 1 \end{pmatrix} . \qquad (7.2.4/24)$$

The elements $a_1$ of the matrix $\underline{A}$ are the coefficients of the denominator $N(s)$ of the transfer function $G(s)$ that means

$$\prod_{l=1}^{q} (s-\lambda_1) = \sum_{l=0}^{q} a_1 s^l \quad . \qquad (7.2.4/25)$$

From eq. (7.2.4/22) we see that a further contribution is added to the output $f(t)$ from eq. (7.2.4/5). This contribution is independent on the switching instants because it is only evoked by the initial conditions which cannot be influenced by the input function $u(t)$. Therefore we can use the relation (7.2.4/16) and must extend it

by this contribution caused by the initial conditions. But this is an easy task. It
is only necessary to determine the inverse Laplace transform of

$$\mathcal{L}\{f_{in}(t)\} = \frac{\underline{s}^T \underline{A} \ \underline{f}(0-)}{N(s)} \ .$$ 

(7.2.4/26)

The numerator $C(s)$ of this expression is a polynomial of degree $q-1$, that means

$$C(s) = \underline{s}^T \underline{A} \ \underline{f}(0-) = c_q s^{q-1} + \ldots + c_1 s + c_0 \ .$$ 

(7.2.4/27)

Because of the assumptions concerning the eigenvalues of the mathematical model of
the transfer system described by the transfer function $G(s)$ it is possible to determine the partial fraction expansion of $C(s)/N(s)$ in the form

$$\mathcal{L}\{f_{in}(t)\} = \frac{C(s)}{N(s)} = \sum_{l=1}^{q} \frac{\tilde{A}_l}{s-\lambda_l}$$

(7.2.4/28)

where the coefficients $\tilde{A}_l$ are given by

$$\tilde{A}_l = \frac{C(\lambda_l)}{\displaystyle\prod_{\substack{k=1 \\ k \neq l}}^{q} (\lambda_l - \lambda_k)} \ .$$

(7.2.4/29)

The inverse Laplace transform of $C(s)/N(s)$ from eq. (7.2.4/28) is given by

$$f_{in}(t) = \sum_{l=1}^{q} \tilde{A}_l e^{\lambda_l t} \sigma(t).$$

(7.2.3/30)

This is the supplementary contribution caused by the initial conditions $\underline{f}(0-)$ different from the null vector. It has to be added to the output $f(t)$ from eq.
(7.2.4/16). Then we obtain for the output $f(t)$ in the presence of initial conditions
different from zero

$$f(t) = \sum_{l=1}^{q} \tilde{A}_l e^{\lambda_l t} \sigma(t) + u_1(r_0 + \sum_{l=1}^{q} r_l e^{\lambda_l t})\sigma(t)$$

$$+ (u_1-u_2) \sum_{j=1}^{n} (-1)^j (r_0 + \sum_{l=1}^{q} r_l e^{\lambda_l (t - \sum_{k=1}^{j} x_k^2)})\sigma(t-\sum_{k=1}^{j} x_k^2) \ .$$

(7.2.4/31)

The derivatives of the functions $f_i$ which result from eq. (7.2.4/31) are given by eq.
(7.2.4/17) because the contribution in the output $f(t)$ evoked by the initial conditions is independent on the switching instants.

With the eqs. (7.2.4/31) respectively (7.2.4/16) and (7.2.4/17) we have a set of
well appropriate analytical formulas at our disposal with which we can utilize one of
the presented methods for the solution of nonlinear least squares problems. As result
of these numerical methods we obtain a parameter vector $\underline{x}^*$. The switching instants $T_j$
of the control function $u(t)$ we are interested in are determined with the help of eq.
(7.2.4/15) from the components of this parameter vector $\underline{x}^*$.

## 7.3 Numerical examples

In the following subsections we present some numerical examples which show the performance of Hartley's method and of several versions of the proposed new method for the solution of nonlinear least squares problems. The examples which follow refer to the problems formulated and studied in the preceding sections.

### 7.3.1 Utilized methods

We consider six different methods for the solution of the resulting nonlinear least squares problems. These methods are denotated by method I to VI and are briefly described in the sequel.

**Method I**

It is Hartley's method where the direction vector $\underline{r}^k$ from eq. (3.2/6) is determined by the aid of a so-called singular value decomposition of the matrix $\underline{D}^k$ [9]. It is advantageous to use such a decomposition because the matrix $(\underline{D}^k)^T\underline{D}^k$ is in general ill-conditioned so that the inverse matrix may be very inaccurate, if we do not use a numerical method which pays attention to this property. The singular value decomposition is an appropriate tool (see Appendix F). In order to perform it we need a numerical method. Such an efficient method is described in [10]. The subroutine given there is written in ALGOL 60. Because all other subroutines for the solution of the nonlinear least squares problems considered in this chapter are written in FORTRAN IV this subroutine has been translated to FORTRAN IV.

**Method II**

This method is only applicable if the functions $f_i(\underline{x})$ depend linearly on some parameters $x_j$. As mentioned in chapter 6 we can divide the nonlinear least squares problem into a linear and a nonlinear one. One possible division is the following one: In the first stage of the iteration we treat the nonlinear least squares problem as a linear one and in the following stages we treat it as a linear one if the first vector $\underline{d}_j^k$ which we select according to eq. (4.3/2) belongs to a parameter $x_j$ on which the functions $f_i(\underline{x})$ depend linearly. In all other cases we solve the complete nonlinear least squares problem. This treatment of the nonlinear least squares problem as a linear and a nonlinear one shall be called "switching between a linear and nonlinear fit". This strategic element is also used in connexion with the methods IV and VI. Naturally it is possible to use other criteria than the just described one for the switching between a linear and nonlinear fit. But here we only utilize the above-mentioned criterion.

In method II we always use all vectors $\underline{d}_j^k$ for the determination of the direction vector $\underline{r}^k$. No selection of the vectors $\underline{d}_j^k$ is performed. In order to avoid the direct inversion of the matrix $(\underline{D}^k)^T\underline{D}^k$ we use the singular value decomposition of the matrix $\underline{D}^k$ (see description of method I).

Because of the similarity of this method to Hartley's method (method I) we also call it modified Hartley's method.

Method III

Here we use the criterion "maximal change of the slope" for the selection of the vectors $\underline{d}_j^k$ for the building up of the matrix $\underline{D}_1^k$ (see subsection 4.4.1). The direction vector $\underline{r}^k$ is determined as described in section 4.3.

Method IV

Here we also use the criterion "maximal change of the slope" for the selection of the vectors $\underline{d}_j^k$ for the building up of the matrix $\underline{D}_1^k$ (see subsection 4.4.1). Moreover we use the strategic element "switching between a linear and nonlinear fit". Therefore method IV is only applicable if the functions $f_i(\underline{x})$ depend linearly on some parameters $x_j$. The direction vector $\underline{r}^k$ is determined as described in section 4.3.

Method V

It is equivalent to method III with the exception that the criterion "maximal angle" (see subsection 4.4.2) is utilized instead of the criterion "maximal change of the slope" for the selection of the vectors $\underline{d}_j^k$ for the building up of the matrix $\underline{D}_1^k$.

Method VI

It is equivalent to method IV with the exception that we use the criterion "maximal angle" (see subsection 4.4.2) instead of the criterion "maximal change of the slope" for the selection of the vectors $\underline{d}_j^k$ for the building up of the matrix $\underline{D}_1^k$.

In all six methods the determination of the step-length factor $\alpha_{min}$ is performed by the procedure described in section 4.7. If we have to solve a linear least squares problem no determination of the step-length factor $\alpha_{min}$ is performed but it is set to 1 according to eq. (5/10). In the methods III to VI a vector $\underline{d}_j^k$ is eliminated from the selection procedure for the building up of the matrix $\underline{D}_1^k$ if it yields an upper bound of the condition number $\varkappa_H$ from eq. (4.5/27) which is greater than

$$5. \ 10^{n/2} \tag{7.3.1/1}$$

whereby n is the number of parameters to be determined. If we obtain a step-length factor $\alpha_{min}$ which is greater than 0.1 the value from eq. (7.3.1/1) for the upper bound of the condition number $\varkappa_H$ remains unchanged. If $\alpha_{min}$ is less than 0.1 the upper bound from eq. (7.3.1/1) is reduced by the factor 5 for the selection of the vectors $\underline{d}_j^k$ in the next stage of the iteration. If in this next stage the step-length factor $\alpha_{min}$ is greater than 0.1 the upper bound for the condition number $\varkappa_H$ is reset to the value from eq. (7.3.1/1). Otherwise it is reduced by the factor 5 once more. Possibly this reduction is performed several times in successive stages of the iteration. Numerical investigations have shown that this processing is very efficient concerning a fast convergence of the methods III to VI.

For the finishing of the iteration the two scalars $\varepsilon_1$ and $\varepsilon_2$ from eqs. (4.6/2) respectively (4.6/4) are decisive.

We choose

$$\varepsilon_1 = 0.001 .$$ (7.3.1/2)

This value of $\varepsilon_1$ guarantees that the absolute value of the difference between the theoretical value 90° and the angles between all column vectors $\underline{d}_j^k$ and the error vector $\underline{e}(\underline{x}^k)$ is less than 0.06°.

The scalar $\varepsilon_2$ is chosen as

$$\varepsilon_2 = 2.5 \ 10^{-7} m$$ (7.3.1/3)

whereby m is the number of given fixed values $y_i$.

The selection of the vectors $\underline{d}_j^k$ for the building up of the matrix $\underline{D}_1^k$ and the simultaneous determination of the direction vector $\underline{r}^k$ are performed in double precision floating point arithmetic when applying methods III to VI. The singular value decomposition and the determination of the direction vector $\underline{r}^k$ in the methods I and II are performed in single precision floating point arithmetic. All other operations in all methods are also performed in single precision floating point arithmetic.

It is to remark that all programs are written in FORTRAN IV and that all numerical calculations were performed on a digital computer IBM 370/158 under OS/VS2. The utilized compiler was the FORTRAN H EXTENDED compiler.

## 7.3.2 Example 1

This example deals with the simplification of a given mathematical model. This problem is to be solved by the aid of a formulation as a nonlinear least squares problem. We demand that the step response of the given mathematical model is fitted as well as possible in the sense of a least sum of squares by the step response of a linear time-invariant mathematical model. The mathematical model - considered in the sequel - is taken from [5]. It is described by a transfer function

$$G(s) = \frac{4000+1000s}{1000+142s+16.9s^2+0.819s^3} .$$ (7.3.2/1)

Eq. (7.3.2/1) means that the given mathematical model is linear and time-invariant. The step response of this model is represented in figure 7/13.

As figure 7/13 shows the step response is a damped oscillation so that it is obvious to describe the step response of the simplified mathematical model by

$$f(t) = A_0 + (B_1\cos\omega_1 t + C_1\sin\omega_1 t)e^{-\delta_1 t} .$$ (7.3.2/2)

Comparing eq. (7.3.2/2) with the general representation (7.2.2/3) for the step response of a linear time-invariant transfer system we find

$$n_r = 0$$ (7.3.2/3)

and

$$n_k = 1 \quad . \tag{7.3.2/4}$$

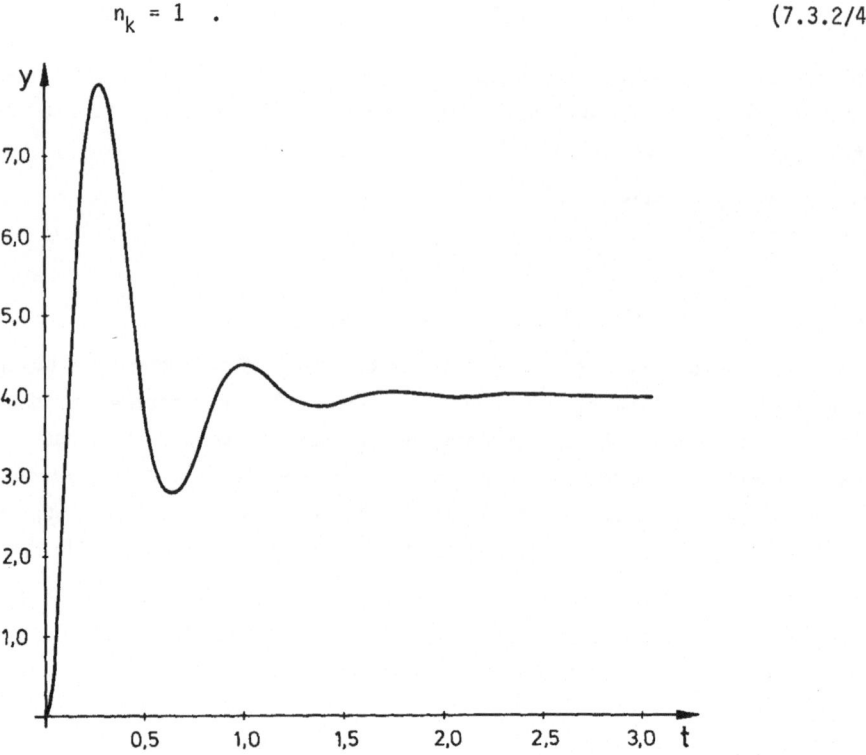

Fig. 7/13: Step response of the mathematical model (7.3.2/1).

From figure 7/13 respectively from eq. (7.3.2/1) we see that the step response is continuous at t = 0. We demand that the step response of the simplified mathematical model has the same property. Therefore we eliminate one parameter from eq. (7.3.2/2), for instance $A_0$, as described in subsection 7.2.2. So we only have four parameters which have to be determined, namely

$$\underline{x} = (C_1, B_1, \omega_1, \delta_1)^T \quad . \tag{7.3.2/5}$$

It is easy to verify that the given mathematical model (7.3.2/1) is stable. We demand that the simplified mathematical model also possesses this property. For that purpose we introduce the parameter $\delta_1'$ from eq. (7.2.2/21) instead of $\delta_1$ in eq. (7.3.2/2).

In order to obtain a nonlinear least squares problem we have to determine the given fixed values $y_i$, i = 1, ..., m of the step response for the system (7.3.2/1). That means we must prescribe the points of time $t_i$, i = 1, ..., m for which we want to achieve a good fit of the step response. In the sequel we consider two cases. In case 1 the points of time are given by

$$t_i = 0.02 \ (i-1), \ i = 1, \ \ldots, \ 101 \quad . \tag{7.3.2/6}$$

In case 2 they are given by

$$t_i = 0.02 \ (i-1), \ i = 1, \ldots, 151 \ . \tag{7.3.2/7}$$

The points of time are equidistant in both cases. In case 1 we have m = 101 given values, in case 2 we have m = 151 given values. In case 1 the points of time range from 0 to 2, in case 2 they range from 0 to 3. Considering figure 7/13 we state that we have more values from the stationary part of the step response in case 2 than in case 1. Therefore we can expect that the Bode gain of the simplified mathematical model in case 2 will be closer to the Bode gain of the given mathematical model than in case 1.

For the application of one of the numerical methods for the solution of the resulting nonlinear least squares problem we need the matrix $\underline{P}$ from eq. (2/11). Here we consider the case that

$$\underline{P} = \underline{I} \tag{7.3.2/8}$$

holds that means all components of the error vector have the same influence on the sum of squares.

For the above-described problem of the fit of the step response we want to study the performance of the methods I to VI when starting these methods with different starting points. For the comparison of these methods we utilize the criteria "number of iteration steps" and "computation time in seconds". When utilizing the criterion "number of iteration steps" we count the steps necessary to reach the point $\underline{x}^*$ from a given starting point $\underline{x}^0$. When utilizing the criterion "computation time in seconds" we measure the time necessary for an iterative method to compute the point $\underline{x}^*$ from a starting point $\underline{x}^0$. For ten different starting points the results are represented in the tables 1 and 2 for the points of time from eq. (7.3.2/6) and in the tables 3 and 4 for the points of time from eq. (7.3.2/7).

| Starting point $\underline{x}^0$ | Number of iteration steps | | | | | |
|---|---|---|---|---|---|---|
| | Method I | Method II | Method III | Method IV | Method V | Method VI |
| 3, 1, 1, 2 | — | 19 | 10 | 17 | 16 | 11 |
| 3, 1, 1, 4 | 9 | 10 | 9 | 10 | 9 | 10 |
| 3, 1, 1, 6 | 6 | 8 | 6 | 6 | 6 | 6 |
| 3, 1, 1, 8 | >25 | 6 | 12 | 6 | 9 | 6 |
| 3, 1, 1, 10 | — | 7 | 7 | 7 | 9 | 7 |
| 3, 1, 1, 12 | 10 | 8 | 14 | 8 | 20 | 8 |
| 3, 1, 1, 14 | 17 | 9 | 14 | 9 | 14 | 9 |
| 3, 1, 1, 16 | 13 | 9 | 14 | 9 | 14 | 9 |
| 3, 1, 1, 18 | >25 | 10 | 22 | 10 | >25 | 10 |
| 3, 1, 1, 20 | 12 | 8 | 13 | 8 | 13 | 8 |

Table 1: Number of iteration steps in the case m = 101.

| Starting point $\underline{x}^0$ | Computation time in seconds | | | | | |
|---|---|---|---|---|---|---|
| | Method I | Method II | Method III | Method IV | Method V | Method VI |
| 3, 1, 1, 2 | — | 8.2 | 5.7 | 8.3 | 9.5 | 4.5 |
| 3, 1, 1, 4 | 4.1 | 3.7 | 5.5 | 4.9 | 5.6 | 4.6 |
| 3, 1, 1, 6 | 2.8 | 2.4 | 3.6 | 2.6 | 3.9 | 2.7 |
| 3, 1, 1, 8 | >10.8 | 2.0 | 6.8 | 2.4 | 5.5 | 2.3 |
| 3, 1, 1, 10 | — | 2.6 | 3.9 | 3.1 | 5.7 | 3.0 |
| 3, 1, 1, 12 | 4.7 | 2.9 | 7.8 | 3.6 | 12.3 | 3.6 |
| 3, 1, 1, 14 | 7.2 | 3.4 | 7.6 | 4.3 | 8.5 | 4.3 |
| 3, 1, 1, 16 | 5.9 | 3.7 | 8.2 | 4.7 | 8.8 | 5.1 |
| 3, 1, 1, 18 | >10.3 | 3.8 | 12 | 5.0 | >13.2 | 5.3 |
| 3, 1, 1, 20 | 5.6 | 3.1 | 7.6 | 3.8 | 7.9 | 3.9 |

Table 2: Computation time in seconds in the case m = 101.

| Starting point $\underline{x}^0$ | Number of iteration steps | | | | | |
|---|---|---|---|---|---|---|
| | Method I | Method II | Method III | Method IV | Method V | Method VI |
| 3, 1, 1, 2 | — | >25 | 15 | 18 | 11 | 11 |
| 3, 1, 1, 4 | 9 | 10 | 9 | 10 | 9 | 10 |
| 3, 1, 1, 6 | 7 | 8 | 7 | 7 | 7 | 7 |
| 3, 1, 1, 8 | >25 | 6 | 11 | 6 | 9 | 6 |
| 3, 1, 1, 10 | — | 7 | 7 | 6 | 9 | 6 |
| 3, 1, 1, 12 | — | 9 | 23 | 7 | 23 | 7 |
| 3, 1, 1, 14 | >25 | 10 | 15 | 9 | 14 | 9 |
| 3, 1, 1, 16 | 10 | 9 | 10 | 9 | 10 | 9 |
| 3, 1, 1, 18 | >25 | 10 | 16 | 10 | 14 | 10 |
| 3, 1, 1, 20 | 17 | 8 | 15 | 8 | 14 | 8 |

Table 3: Number of iteration steps in the case m = 151.

| Starting point $\underline{x}^0$ | Computation time in seconds | | | | | |
|---|---|---|---|---|---|---|
| | Method I | Method II | Method III | Method IV | Method V | Method VI |
| 3, 1, 1, 2 | — | >14.0 | 11.6 | 11.7 | 9.5 | 9.2 |
| 3, 1, 1, 4 | 5.6 | 5.3 | 7.6 | 7.2 | 7.7 | 7.2 |
| 3, 1, 1, 6 | 4.4 | 3.6 | 6.2 | 4.7 | 6.2 | 4.7 |
| 3, 1, 1, 8 | >14.4 | 3.0 | 8.6 | 3.6 | 7.5 | 3.7 |
| 3, 1, 1, 10 | — | 3.5 | 5.8 | 3.9 | 7.7 | 4.2 |
| 3, 1, 1, 12 | — | 4.7 | 17.7 | 4.9 | 19.0 | 4.9 |
| 3, 1, 1, 14 | >14.8 | 5.6 | 11.6 | 6.4 | 11.3 | 6.7 |
| 3, 1, 1, 16 | 6.5 | 5.3 | 8.6 | 7.2 | 8.7 | 7.6 |
| 3, 1, 1, 18 | >14.8 | 4.8 | 13.1 | 6.8 | 12.0 | 6.9 |
| 3, 1, 1, 20 | 11.0 | 4.7 | 12.2 | 6.7 | 9.4 | 6.4 |

Table 4: Computation time in seconds in the case m = 151.

The notion "-" means that the method cannot be applied for the necessary inversion of the matrix $(\underline{D}^k)^T \underline{D}^k$ cannot be performed because the matrix is singular or nearly singular in one stage of the iteration. The notion ">25" means that the method is applicable in the course of the first 25 iterations but that it does not yield a point $\underline{x}^*$ for which the gradient of the sum of squares vanishes that means for which the inequality (4.6/2) is satisfied with $\varepsilon_1$ from eq. (7.3.1/2). According to this the meaning of the notions is in the tables "computation time in seconds".

Inspecting the tables 1 to 4 we recognize that method I (Hartley's method) is not appropriate for the solution of the resulting nonlinear least squares problem because it is often not possible to determine the direction vector $\underline{r}^k$ or because it does not converge in 25 iteration steps. In comparison with method I, method II yields an essential improvement. For m = 101 it always converges. For m = 151 it converges for 9 out of 10 starting points. Comparing method II with method I for those cases in which both methods yield the point $\underline{x}^*$ we state that method II needs a shorter computation time than method I. Concerning the number of iteration steps method II has not always an advantage compared with method I. We recognize that the performance of method II is pretty sufficient. But it is to emphasize that method II has a principal drawback namely that it cannot be guaranteed that a determination of the direction vector $\underline{r}^k$ is always possible. Moreover method II is only applicable if the function $\underline{f}(\underline{x})$ depends linearly on some parameters - a situation which is not always present (compare the problems described in subsections 7.2.1 and 7.2.4). Methods III to VI do not show this drawback. They are always applicable for the solution of nonlinear least squares problems. With the exception of one case they always yield the point $\underline{x}^*$ within 25 iteration steps.

Comparing method III with method IV respectively method V with method VI we see that the strategic element "switching between a linear and nonlinear fit" in general reduces the number of iteration steps. The effectiveness of this strategic element becomes evident if we compare the computation times for the methods III and IV respectively for the methods V and VI. They are drastically reduced to some extent. Therefore it is intelligent to use the strategic element "switching between a linear and nonlinear fit" in combination with the proposed method for the solution of nonlinear least squares problems if the function $\underline{f}(\underline{x})$ depends linearly on some parameters. The application of this strategic element is advantageous in connection with the selection of the vectors $\underline{d}_j^k$ for the building up of the matrix $\underline{D}_1^k$ according to the criteria "maximal change of the slope" and "maximal angle". If methods I and II are applicable - which is not sure a priori - this strategic element also yields a shorter computation time. Therefore we can state that for all problems in which the function $\underline{f}(\underline{x})$ depends linearly on some parameters the strategic element "switching between a linear and nonlinear fit" should be used.

Comparing method III with method V respectively method IV with method VI we can state that none of the two criteria "maximal change of the slope" and "maximal angle" has a clear advantage for this problem under consideration here. The situation is quite different if we consider problems in which the function $\underline{f}(\underline{x})$ depends nonlinearly on all parameters. In these cases the proposed method with the selection of the vectors $\underline{d}_j^k$ according to the criterion "maximal change of the slope" is better appropriate than the method with the selection of the vectors $\underline{d}_j^k$ according to the criterion "maximal angle". Therefore in these cases we only consider the methods I and III.

Returning to the tables 1 to 4 we see that the starting point is decisive whether a method converges and in case how many iteration steps are necessary. As measure how "good" a starting point is we can take the ratio of the sum of squares for the starting point $\underline{x}^0$ and of the sum of squares for $\underline{x}*$. These ratios are represented in table 5 for the different starting points considered here. We see that they increase in the main if the last component of the starting vector $\underline{x}^0$ is increased. Looking at table 1 and 3 we recognize that the methods I and II converge for those starting points for which this ratio is pretty great but not for those for which it is smaller than for others. Hence it follows for this example that the ratio $Q(\underline{x}^0)/Q(\underline{x}*)$ is not solely decisive for the goodness of a starting point rather the shape of the function to be minimized will play an important part.

| Starting point $\underline{x}^0$ | $Q(\underline{x}^0)/Q(\underline{x}*)$ | |
|---|---|---|
| | m = 101 | m = 151 |
| 3, 1, 1, 2 | 302.94 | 429.63 |
| 3, 1, 1, 4 | 308.63 | 435.13 |
| 3, 1, 1, 6 | 314.46 | 440.77 |
| 3, 1, 1, 8 | 320.19 | 446.37 |
| 3, 1, 1, 10 | 325.13 | 451.16 |
| 3, 1, 1, 12 | 327.86 | 453.80 |
| 3, 1, 1, 14 | 328.65 | 454.60 |
| 3, 1, 1, 16 | 328.60 | 454.53 |
| 3, 1, 1, 18 | 328.28 | 454.21 |
| 3, 1, 1, 20 | 327.91 | 453.87 |

Table 5: Ratios $Q(\underline{x}^0)/Q(\underline{x}*)$ for different starting points $\underline{x}^0$.

The vector $\underline{x}$ for which the sum of squares becomes minimal is

$$\underline{x}_1^* = (\pm 3.842, -4.092, \pm 8.128, 2.034)^T \qquad (7.3.2/9)$$

for m = 101 and

$$\underline{x}_2^* = (\pm 3.904, -4.062, \pm 8.098, 2.038)^T \qquad (7.3.2/10)$$

for m = 151. In each case we obtain two vectors $\underline{x}_1^*$ respectively $\underline{x}_2^*$ because the trigonometric functions sine and cosine which appear in eq. (7.3.2/2) are odd respectively even. From the eqs. (7.3.2/9) and (7.3.2/10) we derive the transfer functions $\hat{G}_1(s)$ and $\hat{G}_2(s)$ we are interested in, namely

$$\hat{G}_1(s) = \frac{287.236+39.553s}{70.201+4.068+s^2} \qquad (7.3.2/11)$$

and

$$\hat{G}_2(s) = \frac{283.243+39.896s}{69.737+4.077s+s^2} \quad . \qquad (7.3.2/12)$$

We recognize that the parameter vectors $\underline{x}_1^*$ and $\underline{x}_2^*$ and with them the transfer functions do not differ very much. The eigenvalues of the two simplified mathematical models are practically identical as can be seen by comparing the last two components of the vectors $\underline{x}_1^*$ and $\underline{x}_2^*$ from the eqs. (7.3.2/9) and (7.3.2/10) which are the imaginary and real part of the eigenvalues. The Bode gains differ a little. For m = 101 the Bode gain is

$$V_1 = 4.092 \qquad (7.3.2/13)$$

and for m = 151 it is

$$V_2 = 4.062 \quad . \qquad (7.3.2/14)$$

The Bode gain of the given mathematical model is V = 4, that means the relative error of the Bode gain $V_1$ is 2.3 % and the relative error of the Bode gain $V_2$ is 1.5 %. But this is the expected result because in the case m = 151 we have taken more values from the stationary part of the step response for the determination of the simplified mathematical model.

The step responses of the given mathematical model and the simplified mathematical models are shown in the figures 7/14 and 7/15. We recognize a satisfying fit of the given mathematical model.

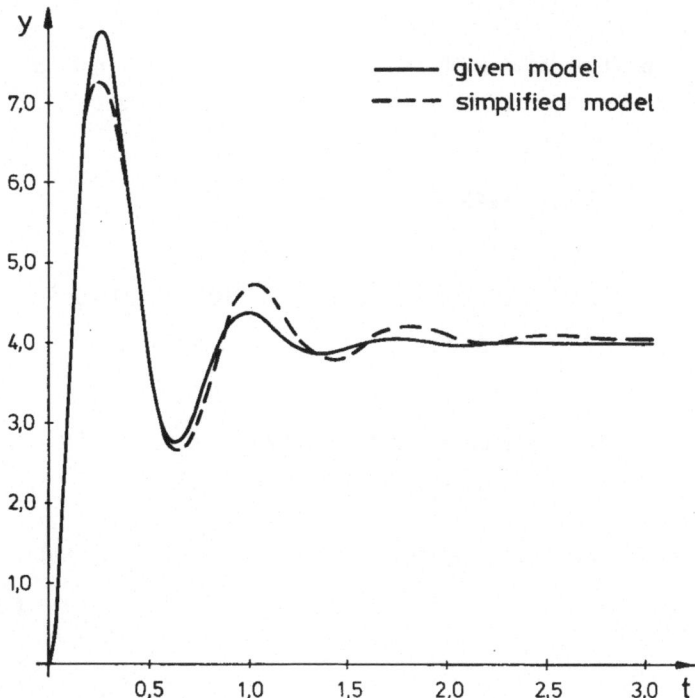

Fig. 7/14: Step responses of the given and simplified mathematical model in the case
m = 101.

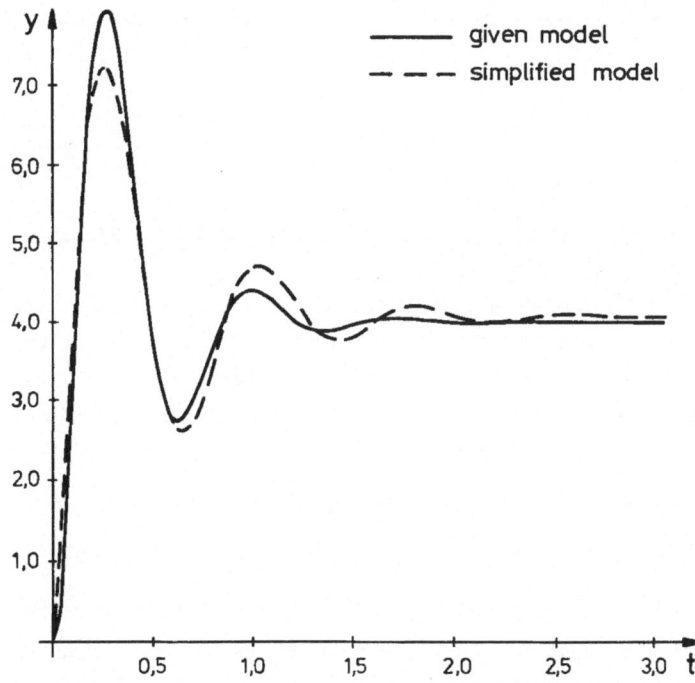

Fig. 7/15: Step responses of the given and simplified mathematical model in the case
m = 151.

### 7.3.3 Example 2

The second example also deals with the problem of the determination of a simplified mathematical model for a given mathematical model. The mathematical model considered in the sequel is taken from [11]. It is given by

$$G(s) = \frac{Z(s)}{N(s)} \qquad (7.3.3/1)$$

with

$$Z(s) = 194480 + 482964s + 511812s^2 + 278376s^3 +$$
$$+ 82402s^4 + 13285s^5 + 1086s^6 + 35s^7 \qquad (7.3.3/2)$$

and

$$N(s) = 9600 + 28880s + 37492s^2 + 27470s^3 + 11870s^4 +$$
$$+ 3017s^5 + 437s^6 + 33s^7 + s^8 \quad . \qquad (7.3.3/3)$$

The pole-zero map for this mathematical model is represented in figure 7/16.

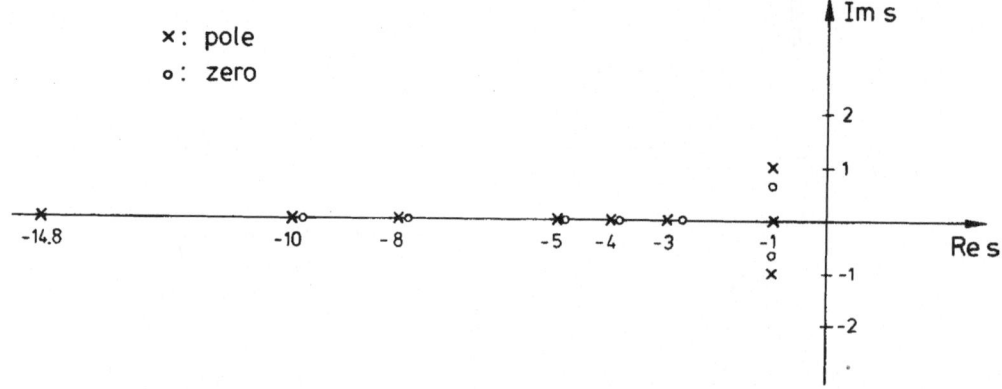

Fig. 7/16: Pole-zero map of the given mathematical model (7.3.3/1).

We look for a simplified mathematical model of the given mathematical model, that means we look for a transfer function the order of which is less than that of the given model. This model shall be derived by a formulation of the problem as a nonlinear least squares problem. To do this we have several possibilities. Here we investigate the following three cases:

Case 1: Fit of the step response of the given mathematical model

Case 2: Fit of the impulse response of the given mathematical model

Case 3: Fit of the magnitude and phase angle plot of the given mathematical model.

In this connection we want to study the performance of the methods I to IV in the cases 1 and 2 and of the methods I and III in the case 3. The methods V and VI are not considered in the sequel because their performance is worse than that of the methods III and IV especially in case 3. The methods II and IV are not applicable in case 3 because the functions $f_i(\underline{x})$ do not depend linearly on some parameters. Moreover we want to investigate whether the mathematical models obtained in the three cases differ.

At first we consider the cases 1 and 2, that means the fit of the step and impulse response. For that purpose it is necessary to compute these values for certain points of time from the given mathematical model. We select 200 equidistant points of time in the interval (0,4]. The values of the step respectively impulse response of the given mathematical model are the given fixed values $y_i$ which shall be fitted by the step respectively impulse response of a linear time-invariant transfer system. For such a system the step response is given by eq. (7.2.2/3) and the impulse response by eq. (7.2.3/2). For the step response we demand that it is continuous at t = 0. In both cases we demand that the resulting linear transfer system is stable. As matrix $\underline{P}$ from eq. (2/11) we use the identity matrix $\underline{I}$. The fit of the given step respectively impulse response shall be done by different simplified mathematical models (comp. the results given in [11]). These models are characterized by different values for $n_r$ and $n_k$ (comp. eqs. (7.2.2/3) and (7.2.3/2)).

For the fit of the step response some important informations are represented in table 6. There the starting points for different mathematical models, the least sums of squares and the ratios $Q(\underline{x}^0)/Q(\underline{x}^*)$ can be found. For $n_r$ = 2 we have chosen two different starting points. We recognize that we obtain two different sums of squares. That means that the function $Q(\underline{x})$ has at least two minima for $n_r$ = 2. Which of these minima is reached depends on the starting point. Inspecting the two models with $n_r$ = 2 it turns out that one of these models leads to the same value of the least sum of squares as the model with $n_r$ = 1. That means this model with $n_r$ = 2 is equivalent to the model with $n_r$ = 1. In the pole-zero map the zero just compensates one pole so that one pole only remains which is just the model with $n_r$ = 1. For the other model with $n_r$ = 2 this compensation does not take place so that we really obtain a model of second order.

| Mathematical model | Starting point $\underline{x}^0$ | $Q(\underline{x}^*)$ | $Q(\underline{x}^0)/Q(\underline{x}^*)$ |
|---|---|---|---|
| $n_r$ = 1 | 1, 2 | 21.8 | 3383.8 |
| $n_r$ = 2 | 1, 1, 1, 2 | 21.8 | 3669.0 |
| $n_r$ = 2 | -31, 11, 2.3, 3.9 | 8.4 | 3.7 |
| $n_k$ = 1 | 1, 1, 0.9, 1.45 | 0.285 | 257222.6 |

Table 6: Results of the fit of the step response for different mathematical models.

Considering the four mathematical models we see that the model with $n_k = 1$ leads to the least sum of squares. Hence it follows that this model is the best in the sense of a least sum of squares if we only have the choice between these four models. The value $Q(\underline{x}*) = 0.285$ means a good fit of this given step response. Therefore we can state that the model with $n_k = 1$ is a well appropriate simplified mathematical model of the given model. The same is true if we consider table 7 where the corresponding informations are represented for the fit of an impulse response. The model with $n_k = 1$ leads to a least sum of squares of 8.2 which also means a good fit of this given impulse response.

| Mathematical model | Starting point $\underline{x}^0$ | $Q(\underline{x}*)$ | $Q(\underline{x}^0)/Q(\underline{x}*)$ |
|---|---|---|---|
| $n_r = 1$ | 1, 2 | 58.1 | 314.9 |
| $n_r = 2$ | 1, 1, 1, 2 | 58.1 | 292.7 |
| $n_k = 1$ | 1, 1, 0.9, 1.45 | 8.2 | 2184.7 |

Table 7: Results of the fit of the impulse response for different mathematical models.

The performance of the methods I to IV for the solution of the resulting nonlinear least squares problem can be judged by considering the tables 8 and 9. The mathematical models in the tables 6 and 8 respectively in the tables 7 and 9 correspond to each other. In table 8 we find the essential informations for the fit of the step response, in table 9 the corresponding informations for the fit of the impulse response are represented. From these tables we see that the methods I and II do not always yield the desired minimum of the sum of squares. (For the notions "-" and ">" compare subsection 7.3.2. Here we have admitted a maximal number of 100 iteration steps.) The methods III and IV always yield a minimum. We see once more that the strategic element "switching between a linear and nonlinear fit" in general reduces the computation times drastically. Comparing method I with method III respectively method II with method IV for those cases in which they both yield the desired minimum, we recognize that the computation times and numbers of iteration steps are nearly identical. As for example 1 we can state that the ratios $Q(\underline{x}^0)/Q(\underline{x}*)$ are not only decisive for the question whether a method converges or not. For a small ratio $Q(\underline{x}^0)/Q(\underline{x}*)$ in the case $n_r = 2$ with $\underline{x}^0 = (-31, 11, 2.3, 3.9)^T$ (fit of the step response) methods I and II do not converge in 100 iteration steps. On the contrast they converge for greater ratios. (See for instance table 6 for the fit of the step response and table 7 for the fit of the impulse response in the case $n_r = 1$.) In the tables 8 and 9 we have omitted the methods V and VI because they nearly show the same performance as the methods III and IV and because they are worse in the case of the fit of a magnitude and phase angle plot.

| Mathematical model | Method I | | Method II | | Method III | | Method IV | |
|---|---|---|---|---|---|---|---|---|
| | Computation time in seconds | Number of iteration steps | Computation time in seconds | Number of iteration steps | Computation time in seconds | Number of iteration steps | Computation time in seconds | Number of iteration steps |
| $n_r = 1$ | 2.1 | 5 | 0.9 | 4 | 2.1 | 5 | 0.85 | 4 |
| $n_r = 2$ | — | — | >69.5 | >100 | 9.6 | 12 | 9.1 | 16 |
| $n_r = 2$ | >87.7 | >100 | >75.5 | >100 | 47.4 | 72 | 30.5 | 73 |
| $n_k = 1$ | >68.5 | >100 | 4.4 | 8 | 61.9 | 87 | 51.6 | 71 |

Table 8: Performance of the methods I - IV (fit of the step response).

| Mathematical model | Method I | | Method II | | Method III | | Method IV | |
|---|---|---|---|---|---|---|---|---|
| | Computation time in seconds | Number of iteration steps | Computation time in seconds | Number of iteration steps | Computation time in seconds | Number of iteration steps | Computation time in seconds | Number of iteration steps |
| $n_r = 1$ | 1.5 | 4 | 0.9 | 4 | 1.5 | 4 | 0.6 | 3 |
| $n_r = 2$ | 10.2 | 7 | >77.2 | >100 | 10.4 | 14 | 11.5 | 18 |
| $n_k = 1$ | >33 | >100 | 4.5 | 8 | 55.9 | 74 | 32.7 | 45 |

Table 9: Performance of the methods I - IV (fit of the impulse response).

This is the third case for which we want to derive a simplified mathematical model. In order to produce the given fixed values $y_i$ we must prescribe the frequencies for which we want to compute the values of the magnitude and phase angle plot from the given mathematical model. In accordance with [11] we select

$$\omega_1 = 0.01$$
$$\omega_i = 1.099975\, \omega_{i-1}, \quad i = 2, \ldots, 100 . \qquad (7.3.3/4)$$

That means we select 100 frequencies. For each frequency we compute the values of the magnitude and phase angle plot so that we have a total of 200 given fixed values $y_i$. These values $y_i$ are to be fitted by the values of a magnitude and phase angle plot of a linear time-invariant transfer system. Such a system can be described by a transfer function as it is given by eq. (7.2.1/1). The values $\tilde{n}_r$, $\tilde{n}_k$, $n_r$, $n_k$ and $\rho$ characterize the mathematical model. As matrix $\underline{P}$ from eq. (2/11) we also use the identity matrix $\underline{I}$.

We demand that the given mathematical model and the simplified model have the same Bode gain so that V from eq. (7.2.1/1) is no parameter but it is known. Therefore it is admissible that we do not consider the transfer function (7.3.3/1) but the corresponding transfer function which has the Bode gain 1. This is done in the sequel.

For different mathematical models the starting points, the least sums of squares, the ratios $Q(\underline{x}^0)/Q(\underline{x}^*)$, the computation times and the numbers of iteration steps for the methods I and III are represented in table 10. We recognize that method I only yields the desired minimum of the sum of squares for two mathematical models. For the two other mathematical models method I does not yield the minimum in 100 iteration steps. Method III always yields the minimum. Comparing method I with method III for those mathematical models and starting points for which both methods converge we see that the two methods require the same number of iteration steps and almost the same computation times. So none of these methods has an advantage for these starting points. Here it turns out once more that the ratios $Q(\underline{x}^0)/Q(\underline{x}^*)$ are not only decisive for the question whether a method converges or not. Considering the minimal sums of squares for the different mathematical models we recognize that the model characterized by $\tilde{n}_r$ = 1 and $n_k$ = 1 yields the least sum of squares. This is just the same kind of model as in the case of the fit of an impulse or step response. Comparing the models with $n_r$ = 1 and $n_r$ = 2 we state that the least sums of squares are nearly identical. That means the model with $n_r$ = 2 is practically identical with the model $n_r$ = 1. This happens if one of the parameters $\omega_1'$ respectively $\omega_2'$ in the model with $n_r$ = 2 is much greater than the upper bound of the interval from which the interesting frequencies are taken. (Here this upper bound is $\omega_{100}$ = 125.) Just this is done by methods I and III.

| Mathematical model | Starting point $\underline{x}^0$ | $Q(\underline{x}^*)$ | $Q(\underline{x}^0)/Q(\underline{x}^*)$ | Method I | | Method III | |
|---|---|---|---|---|---|---|---|
| | | | | Computation time in seconds | Number of iteration steps | Computation time in seconds | Number of iteration steps |
| $n_r$ = 1 | 0.5 | 235.0 | 132.4 | 2.8 | 8 | 2.5 | 8 |
| $n_r$ = 2 | 0.5, 0.2 | 234.2 | 3331.4 | 3.4 | 7 | 3.4 | 7 |
| $\bar{n}_r$ = 1, $n_r$ = 2 | 0.5, 0.2, 1. | 228.8 | 1741.7 | > 64.6 | >100 | 9.6 | 13 |
| $\tilde{n}_r$ = 1, $n_k$ = 1 | 0.8, 0.2, 1. | 68.8 | 4250.8 | > 69.7 | >100 | 51.9 | 63 |

Table 10: Performance of the methods I and III (fit of the magnitude and phase angle plot).

If we look for the best simplified mathematical model in the sense of a least sum of squares we state that the kind of model is equal independent on the fact whether we take as given fixed values $y_i$ those of a step, an impulse response, or a magnitude and phase angle plot. But the resulting models are not really identical. In order to demonstrate the differences the pole-zero maps for the three cases considered here are shown in figure 7/17. We see that the locations of the poles and the zeros are not identical but that they are very close together. Hence it follows that the three models will nearly show the same dynamic performance.

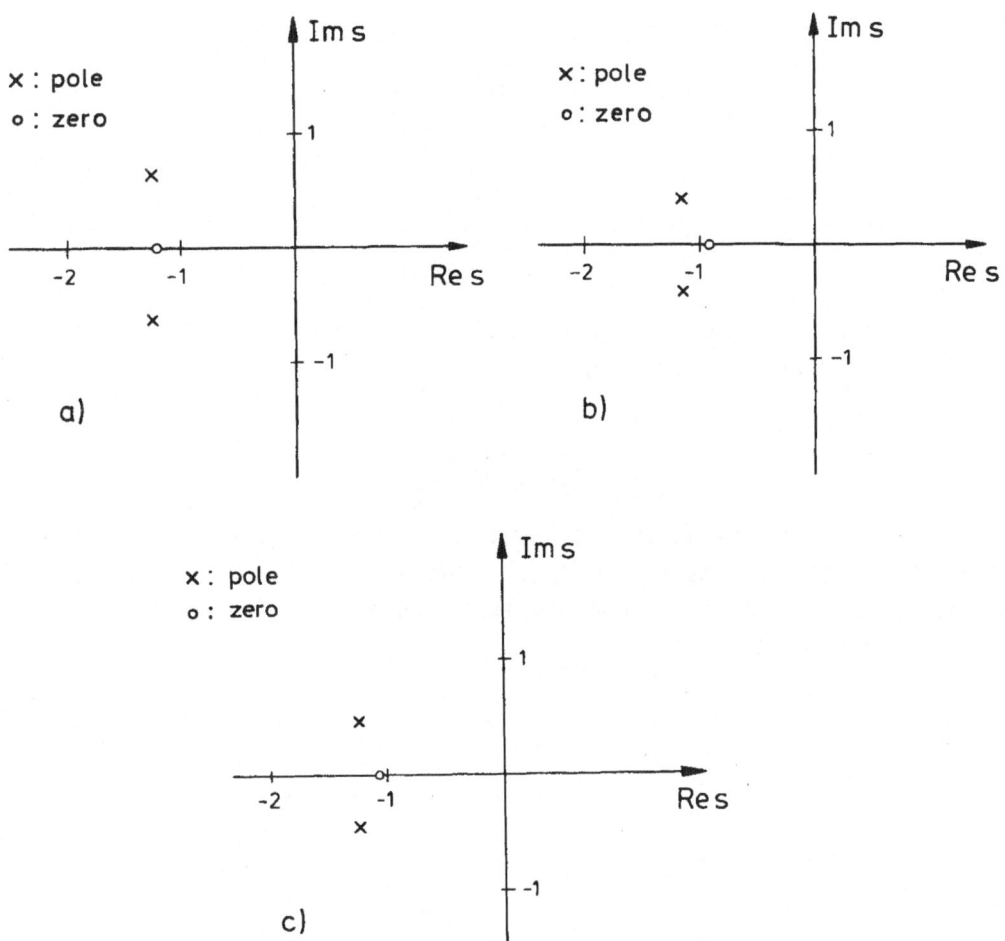

Fig. 7/17: Pole-zero maps of the simplified mathematical models obtained by the
a) fit of the step response
b) fit of the impulse response
c) fit of the magnitude and phase angle plot

## 7.3.4 Example 3

As next example we consider the problem that we want to determine an optimal control function. As shown in subsection 7.2.4 this problem can be solved by means of an appropriate formulation as a nonlinear least squares problem. For the example considered here we assume that the transfer system is described by the transfer function

$$G(s) = \frac{630}{(s-1)(s-1.5)(s-2)(s-5)(s-6)(s-7)} \ . \qquad (7.3.4/1)$$

This transfer function has the general form (7.2.4/3). For this transfer system we demand that its output $f(t)$ shows a desired performance $y(t)$, namely

$$y(t) = \begin{cases} 2/15 \ t \ \text{for} \ 0 \ \le t \le \ 1.5 \\ 0.2 \ \text{for} \ 1.5 \le t \le \ 5 \\ - 22/15 + 1/3 \ t \ \text{for} \ 5 \ \le t \le \ 6.5 \\ 0.7 \ \text{for} \ 6.5 \le t \le \ 8 \\ - 0.9 \ + 0.2 \ t \ \text{for} \ 8 \ \le t \le \ 9.5 \\ 1.0 \ \text{for} \ 9.5 \le t \le 18 \end{cases} \qquad (7.3.4/2)$$

We want to consider the prescribed output $y(t)$ only for discrete points of time $t_i$, i = 1, ..., m. Let these points of time be

$$t_1 = 0.25$$
$$t_i = t_{i-1} + 0.25, \ i = 2, \ ..., \ 72. \qquad (7.3.4/3)$$

That means we have a total of 72 given points of time $t_i$ ranging equidistantly from 0.25 to 18. For these points of time $t_i$ the values of the prescribed output $y(t)$ are calculated from eq. (7.3.4/2). They are the given fixed values $y_i$ which are represented in figure 7/18.

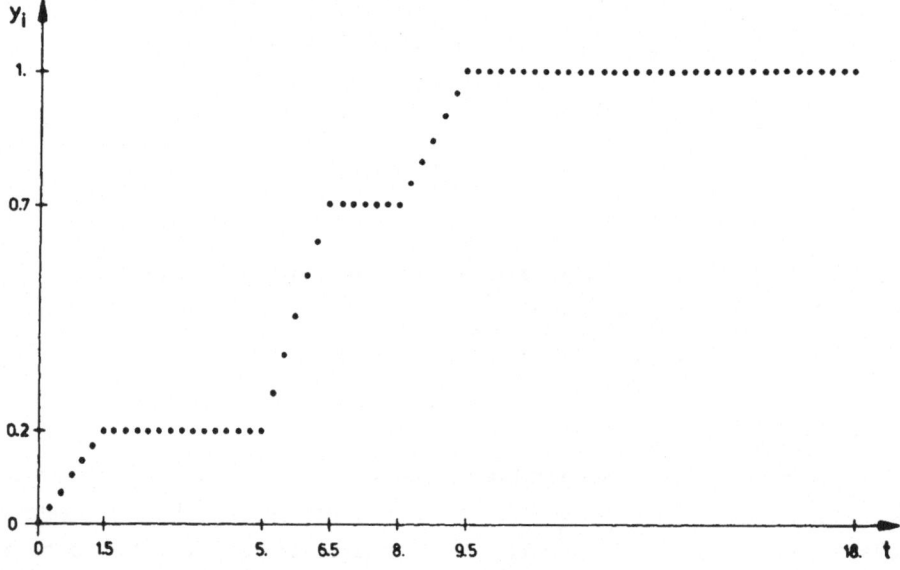

Fig. 7/18: Given values $y_i$ of the prescribed output.

Now we look for a control function u(t) which causes the output f(t) of the given transfer system to show the performance y(t) for the given points of time $t_i$. In order to solve this problem by an appropriate formulation as a nonlinear least squares problem we must parametrize the control function u(t) adequately. As control function u(t) we use the function represented in fig. 7/11. We assume that the values $u_1$ and $u_2$ are known but that the switching instants $T_j$ are still to be determined. That means that the switching instants $T_j$ are the parameters which are to be calculated by the solution of a nonlinear least squares problem. This problem can be formulated as follows: Determine the switching instants $T_j$ so that the given values $y_i$ of the desired output y(t) are fitted as well as possible by the values $f(t_i)$ of the output of the transfer system (7.3.4/1) when applying the control function from fig. 7/11.

We assume that all initial conditions for the transfer system (7.3.4/1) are zero at t = 0. That means that the output f(t) of the transfer system is given by eq. (7.2.4/12) for the control function applied here. As described in subsection 7.2.4 we introduce new parameters instead of the switching instants $T_j$ according to eq. (7.2.4/15). In order to solve the resulting nonlinear least squares problem it is necessary to prescribe the number of switching instants $T_j$ - that means the number n of parameters - and to choose a suitable starting point. In the sequel we consider six different numbers for the switching instants, namely 8, 10, 12, 15, 20, and 25 switching instants. Moreover we consider two cases for the values $u_1$ and $u_2$ of the control function u(t) namely

$$u_1 = 2 , \quad u_2 = 0 \qquad\qquad (7.3.4/4)$$

and

$$u_1 = 2 , \quad u_2 = -1 . \qquad\qquad (7.3.4/5)$$

As matrix $\underline{P}$ from eq. (2/11) we use the identity matrix $\underline{I}$.

In the tables 11 - 16 the starting values $T_j^0$ and the optimal values $T_j^*$ are represented for the different numbers of switching instants considered here. (For the application of the methods I and III the components of the starting point $\underline{x}^0$ are calculated from the values $T_j^0$ according to eq. (7.2.4/15).) The notion u = 2; 0 means the values for $u_1$ and $u_2$ according to eq. (7.3.4/4), the notion u = 2; -1 means the values for $u_1$ and $u_2$ according to eq. (7.3.4/5). As can be expected the optimal switching instants $T_j^*$ depend on the fact whether $u_2$ has the value 0 or -1.

| $T_j^o$ | $T_j^*$ | |
|---|---|---|
| | u=2;0 | u=2;-1 |
| 2 | 0.389 | 1.066 |
| 4 | 4.324 | 3.347 |
| 6 | 5.609 | 5.438 |
| 8 | 7.503 | 6.784 |
| 10 | 9.435 | 9.482 |
| 12 | 11.323 | 10.688 |
| 14 | 13.205 | 13.279 |
| 16 | 15.084 | 14.481 |

Table 11: Starting and optimal values in the case of 8 switching instants.

| $T_j^o$ | $T_j^*$ | |
|---|---|---|
| | u=2;0 | u=2;-1 |
| 2 | 0.383 | 1.027 |
| 4 | 4.193 | 3.227 |
| 5 | 5.356 | 5.132 |
| 6 | 7.076 | 6.378 |
| 8 | 8.645 | 8.533 |
| 10 | 10.072 | 9.487 |
| 11 | 11.477 | 11.420 |
| 13 | 12.881 | 12.366 |
| 14 | 14.285 | 14.296 |
| 15 | 15.684 | 15.239 |

Table 12: Starting and optimal values in the case of 10 switching instants.

| $T_j^o$ | $T_j^*$ | |
|---|---|---|
| | u=2,0 | u=2,-1 |
| 2 | 0.381 | 0.649 |
| 4 | 4.148 | 1.981 |
| 5 | 5.264 | 2.874 |
| 6 | 6.899 | 4.086 |
| 8 | 8.267 | 5.731 |
| 9 | 9.415 | 6.801 |
| 10 | 10.527 | 8.843 |
| 11 | 11.639 | 9.768 |
| 13 | 12.752 | 11.633 |
| 14 | 13.863 | 12.548 |
| 15 | 14.974 | 14.409 |
| 16 | 16.082 | 15.321 |

Table 13: Starting and optimal values in the case of 12 switching instants.

| $T_j^o$ | $T_j^*$ | |
|---|---|---|
| | u=2,0 | u=2,-1 |
| 1 | 0.324 | 0.632 |
| 2 | 2.654 | 1.920 |
| 3 | 2.888 | 2.781 |
| 4 | 4.399 | 3.988 |
| 5 | 5.418 | 5.546 |
| 6 | 6.930 | 6.542 |
| 7 | 8.212 | 8.205 |
| 8 | 9.252 | 8.900 |
| 9 | 10.246 | 10.266 |
| 10 | 11.244 | 10.955 |
| 11 | 12.243 | 12.341 |
| 12 | 13.242 | 13.031 |
| 13 | 14.242 | 14.418 |
| 14 | 15.241 | 15.109 |
| 15 | 16.240 | 16.496 |

Table 14: Starting and optimal values in the case of 15 switching instants.

| $T_j^o$ | $T_j^*$ | |
|---|---|---|
| | $u = 2_i 0$ | $u = 2_i -1$ |
| 0.5 | 0.317 | 0.522 |
| 3.1 | 2.449 | 1.474 |
| 3.3 | 2.649 | 2.066 |
| 4.5 | 4.261 | 2.939 |
| 5.5 | 5.191 | 3.510 |
| 6.5 | 6.354 | 4.213 |
| 7.5 | 6.878 | 5.452 |
| 8.5 | 7.483 | 6.212 |
| 9.0 | 8.450 | 7.114 |
| 10.0 | 9.300 | 7.552 |
| 10.7 | 10.040 | 8.783 |
| 11.4 | 10.776 | 9.387 |
| 12.1 | 11.511 | 10.541 |
| 12.8 | 12.246 | 11.111 |
| 13.5 | 12.980 | 12.252 |
| 14.2 | 13.714 | 12.822 |
| 14.9 | 14.448 | 13.962 |
| 15.6 | 15.181 | 14.531 |
| 16.3 | 15.914 | 15.671 |
| 17.0 | 16.647 | 16.239 |

Table 15: Starting and optimal values in the case of 20 switching instants.

| $T_j^o$ | $T_j^*$ | |
|---|---|---|
| | u = 2,0 | u = 2,-1 |
| 0.5 | 0.309 | 0.508 |
| 2.5 | 2.130 | 1.397 |
| 2.7 | 2.263 | 1.926 |
| 3.1 | 3.127 | 2.688 |
| 3.3 | 3.207 | 3.192 |
| 4.5 | 4.196 | 3.947 |
| 5.5 | 4.903 | 4.878 |
| 5.9 | 5.389 | 5.239 |
| 6.1 | 5.750 | 5.891 |
| 6.8 | 6.784 | 6.537 |
| 7.0 | 7.504 | 7.432 |
| 7.5 | 7.952 | 7.763 |
| 7.7 | 8.660 | 8.746 |
| 8.2 | 9.421 | 9.254 |
| 8.5 | 10.108 | 10.184 |
| 8.7 | 10.773 | 10.636 |
| 9.0 | 11.435 | 11.537 |
| 10.0 | 12.098 | 11.987 |
| 11.0 | 12.761 | 12.887 |
| 12.0 | 13.425 | 13.336 |
| 13.0 | 14.090 | 14.235 |
| 14.0 | 14.755 | 14.684 |
| 15.0 | 15.420 | 15.583 |
| 16.0 | 16.085 | 16.031 |
| 17.0 | 16.753 | 16.937 |

Table 16: Starting and optimal values in the case of 25 switching instants.

For the starting points given in the tables 11 - 16 the resulting nonlinear least squares problems are solved by the methods I and III. The essential informations about the performance of these methods can be found in table 17 for the values $u_1$ and $u_2$ from eq. (7.3.4/4) and in table 18 for the values $u_1$ and $u_2$ from eq. (7.3.4/5).

| Number of switching instants | $Q(\underline{x}^{*})$ | $Q(\underline{x}^{0})/Q(\underline{x}^{*})$ | Method I | | Method III | |
|---|---|---|---|---|---|---|
| | | | Computation time in seconds | Number of iteration steps | Computation time in seconds | Number of iteration steps |
| 8 | 2.1112 | 6.26 | 100 | 23 | 118 | 24 |
| 10 | 0.6708 | 19.77 | 98 | 20 | 148 | 24 |
| 12 | 0.2947 | 41.14 | 136 | 25 | 212 | 27 |
| 15 | 0.1264 | 66.81 | 206 | 29 | 270 | 28 |
| 20 | 0.0322 | 23.29 | 344 | 37 | 485 | 42 |
| 25 | 0.0218 | 60.71 | 432 | 33 | 918 | 68 |

Table 17: Performance of the methods I and III in the case $u_1 = 2$ and $u_2 = 0$.

| Number of switching instants | $Q(\underline{x}^{*})$ | $Q(\underline{x}^{0})/Q(\underline{x}^{*})$ | Method I | | Method III | |
|---|---|---|---|---|---|---|
| | | | Computation time in seconds | Number of iteration steps | Computation time in seconds | Number of iteration steps |
| 8 | 5.2625 | 4.30 | 123 | 27 | 136 | 23 |
| 10 | 2.7605 | 10.21 | 147 | 28 | 179 | 26 |
| 12 | 1.0595 | 9.71 | 307 | 42 | 453 | 48 |
| 15 | 0.3541 | 44.27 | 303 | 39 | 406 | 37 |
| 20 | 0.0727 | 340.25 | — | — | 1324 | 95 |
| 25 | 0.0236 | 1076.35 | — | — | 1436 | 97 |

Table 18: Performance of the methods I and III in the case $u_1 = 2$ and $u_2 = -1$.

Inspecting table 17 we see that the methods I and III always yield the desired minima of the sum of squares. Comparing the number of iteration steps we see that there are no great differences except in the case of 25 switching instants. Comparing the computation times we can state that method I always needs a shorter computation time than method III. Looking at the ratios $Q(\underline{x}^{0})/Q(\underline{x}^{*})$ we recognize that they are always pretty small. Perhaps this is one reason that method I always converges. The other reason is surely that we have chosen "good" starting points for the iteration.

As can be expected the minimal sums of squares decrease if we increase the number of switching instants. The same result follows from table 18. Comparing the minimal sums of squares from the tables 17 and 18 for the different numbers of switching instants we recognize that those from table 17 are always smaller. This result means that the control function u(t) with $u_1$ and $u_2$ from eq. (7.3.4/4) is better appropriate than that with $u_1$ and $u_2$ from eq. (7.3.4/5). We see that the corresponding ratios "minimal sum of squares from table 17" to "minimal sum of squares from table 18" increase in the main if we increase the number of switching instants. That means that the value $u_2$ is not so important if we have a great number of switching instants.

Looking at table 18 we notice that method I is not applicable in the cases of 20 and 25 switching instants. Method III however yields the minimal sums of squares. Comparing method I with method III in the other cases we see that sometimes method III and sometimes method I needs a lower number of iteration steps. Comparing the computation times we state that method I always needs a shorter computation time than method III, but this fact cannot remove the shortcoming that method I sometimes fails. Looking at the ratios $Q(\underline{x}^0)/Q(\underline{x}^*)$ we see that method I fails in those cases in which these ratios are pretty great compared with the other cases. Considering the ratios $Q(\underline{x}^0)/Q(\underline{x}^*)$ in the tables 17 and 18 we notice that they all are of equal order of magnitude except for the two cases in which method I fails. Hence it follows from this example that the ratio $Q(\underline{x}^0)/Q(\underline{x}^*)$ may be a helpful tool for the investigation of the question why method I sometimes fails.

References:

[1] Isermann, R.:
Theoretische Analyse der Dynamik industrieller Prozesse
Zürich: Bibliographisches Institut (1971).

[2] Eykhoff, P.:
System Identification
Parameter and State Estimation
London: John Wiley & Sons, Ltd. (1974).

[3] Unbehauen, H; Göhring, B.; Bauer, B.:
Parameterschätzverfahren zur Systemidentifikation
Methoden der Regelungstechnik
München und Wien: Oldenbourg Verlag (1974).

[4] Bosley, M.J.; Lees, F.P.:
A Survey of Simple Transfer-Function Derivations from High-Order State-Variable Models
Automatica 8, 765 - 775 (1972).

[5] Gwinner, K.:
Vereinfachung von Modellen dynamischer Systeme
Regelungstechnik 10, 325 - 333 (1976).

[6] DiStefano, J.J.; Stubberud, A.R.; Williams, I.J.:
Theory and Problems of Feedback and Control Systems
Schaum's Outline Series
New York: McGraw-Hill, Inc. (1967).

[7] Coddington, E.A.; Levinson, N.:
Theory of Ordinary Differential Equations
New York: McGraw-Hill, Inc. (1955).

[8] Kiendl, H.:
Zur Behandlung von stückweise differenzierbaren Eingangsfunktionen von zeitinvarianten linearen Differentialgleichungen n-ter Ordnung
Archiv für Elektrotechnik 60, 259 - 266 (1978).

[9] Stewart, G.W.:
Introduction to Matrix Computations
New York and London: Academic Press (1973).

[10] Golub, G.H.; Reinsch, C.:
Singular Value Decomposition and Least Squares Solutions
Numer. Math. 14, 403 - 420 (1970).
[11] Luus, R.:
Optimization in Model Reduction
Int. J. of Control 32, 741 - 747 (1980).

## Appendix A

Some properties of the matrix $\underline{D}^T \underline{P} \ \underline{D}$

The mxn-matrix $\underline{D}^k$ is defined by eq. (3.1/12) with $m \geq n$ (compare eq. 2/16). In the sequel we will omit the index k in order to simplify the notation.

For the rank r of the matrix $\underline{D}$ the inequality

$$\text{rank } \underline{D} = r \leq n \tag{A/1}$$

holds, because the rank of a rectangular matrix is at most as great as the less of the two numbers m or n [1].

By virtue of the assumption the nxn-matrix $\underline{P}$ is a real positive definite symmetric matrix. Because of these assumptions the matrix $\underline{P}$ can be represented as

$$\underline{P} = \underline{S}^T \underline{S} \tag{A/2}$$

where the matrix $\underline{S}$ is an upper triangular matrix [1,2]. (The processing of this decomposition on a digital computer makes no difficulties.)

As the matrix $\underline{P}$ is positive definite, it is in particular nonsingular, that means

$$\text{rank } \underline{P} = n. \tag{A/3}$$

Because of eq. (A/2) we have

$$\text{rank } \underline{S} = \text{rank } \underline{S}^T = n. \tag{A/4}$$

Besides the matrix $\underline{D} = (\underline{\bar{d}}_1, \ldots, \underline{\bar{d}}_n)$ we consider the matrix

$$\underline{D} = (\underline{d}_1, \ldots, \underline{d}_n) = \underline{S} \ \underline{\bar{D}}. \tag{A/5}$$

Now the following theorem is valid [1]:

If we have a system of vectors $\underline{\bar{d}}_1, \ldots, \underline{\bar{d}}_n$ with rank r and if we perform a nonsingular transformation $\underline{S} \ \underline{\bar{d}}_j = \underline{d}_j$, $j = 1, \ldots, n$, then the new system $\underline{d}_1, \ldots, \underline{d}_n$ has the same rank r as the old one. If the vectors $\underline{\bar{d}}_j$ are linearly independent in particular (i. e. r = n), so this is also valid for the transformed vectors $\underline{d}_j$.

This theorem means that the rank of the matrix $\underline{D}$ from eq. (A/5) is given by the rank of the matrix $\underline{\bar{D}}$ if the matrix $\underline{S}$ is nonsingular. Therefore we have the identity

$$\text{rank } \underline{D} = \text{rank } \underline{\bar{D}}. \tag{A/6}$$

As $\underline{\bar{D}}$ is a mxn-matrix with $m \geq n$, it is the easiest way to determine the rank of the matrix $\underline{\bar{D}}$ by finding out the maximal number of linearly independent column vectors. If

$$\text{rank } \underline{\bar{D}} = n \tag{A/7}$$

is valid then the matrix $\underline{\underline{D}}$ has n linearly independent row and column vectors.

By utilization of the eqs. (A/2) and (A/5) we can write

$$\underline{\underline{D}}^T \underline{\underline{P}} \; \underline{\underline{D}} = \underline{\underline{D}}^T \underline{\underline{S}}^T \underline{\underline{S}} \; \underline{\underline{D}} = (\underline{\underline{S}} \; \underline{\underline{D}})^T (\underline{\underline{S}} \; \underline{\underline{D}}) = \underline{\underline{D}}^T \underline{\underline{D}}. \tag{A/8}$$

We notice that the matrix $\underline{\underline{D}}^T \underline{\underline{P}} \; \underline{\underline{D}}$ is symmetric because

$$(\underline{\underline{D}}^T \underline{\underline{P}} \; \underline{\underline{D}})^T = (\underline{\underline{D}}^T \underline{\underline{D}})^T = \underline{\underline{D}}^T \underline{\underline{D}} = \underline{\underline{D}}^T \underline{\underline{P}} \; \underline{\underline{D}} \tag{A/9}$$

holds.

Moreover the matrix $\underline{\underline{D}}^T \underline{\underline{P}} \; \underline{\underline{D}}$ is positive semidefinite. To prove this proposition we must show that

$$\underline{z}^T \underline{\underline{D}}^T \underline{\underline{P}} \; \underline{\underline{D}} \; \underline{z} \geq 0 \tag{A/10}$$

holds for any arbitrarily chosen $\underline{z} \in \mathbb{R}^n$.

By utilization of eq. (A/8) we obtain

$$\underline{z}^T \underline{\underline{D}}^T \underline{\underline{P}} \; \underline{\underline{D}} \; \underline{z} = \underline{z}^T \underline{\underline{D}}^T \underline{\underline{D}} \; \underline{z} = (\underline{\underline{D}} \; \underline{z})^T \underline{\underline{D}} \; \underline{z}. \tag{A/11}$$

With the substitution

$$\underline{v} = \underline{\underline{D}} \; \underline{z} \tag{A/12}$$

we find

$$\underline{z}^T \underline{\underline{D}}^T \underline{\underline{P}} \; \underline{\underline{D}} \; \underline{z} = \underline{v}^T \underline{v} = \| \underline{v} \|^2 \geq 0. \tag{A/13}$$

Eq. (A/13) means that the matrix $\underline{\underline{D}}^T \underline{\underline{P}} \; \underline{\underline{D}}$ is positive semidefinite.

Assume that the column vectors of the matrix $\underline{\underline{D}}$ are linearly independent. By inspecting eq. (A/12) we notice that the vector $\underline{v}$ vanishes if and only if $\underline{z} = \underline{0}$ is valid. But this means that the matrix $\underline{\underline{D}}^T \underline{\underline{P}} \; \underline{\underline{D}}$ is positive definite if we have rank $\underline{\underline{D}}$ = rank $\underline{D}$ = n. If the matrix $\underline{\underline{D}}^T \underline{\underline{P}} \; \underline{\underline{D}}$ is positive definite, it is also invertible, i. e. the inverse matrix $(\underline{\underline{D}}^T \underline{\underline{P}} \; \underline{\underline{D}})^{-1}$ exists. This inverse matrix $(\underline{\underline{D}}^T \underline{\underline{P}} \; \underline{\underline{D}})^{-1}$ is also positive definite. To show this we evaluate

$$\underline{z}^T (\underline{\underline{D}}^T \underline{\underline{P}} \; \underline{\underline{D}})^{-1} \underline{z} = \underline{z}^T (\underline{\underline{D}}^T \underline{\underline{P}} \; \underline{\underline{D}})^{-1} (\underline{\underline{D}}^T \underline{\underline{P}} \; \underline{\underline{D}}) \; (\underline{\underline{D}}^T \underline{\underline{P}} \; \underline{\underline{D}})^{-1} \underline{z}$$

$$= \underline{z}^T [(\underline{\underline{D}}^T \underline{\underline{P}} \; \underline{\underline{D}})^{-1}]^T (\underline{\underline{D}}^T \underline{\underline{P}} \; \underline{\underline{D}}) \; (\underline{\underline{D}}^T \underline{\underline{P}} \; \underline{\underline{D}})^{-1} \underline{z}$$

$$= [(\underline{\underline{D}}^T \underline{\underline{P}} \; \underline{\underline{D}})^{-1} \underline{z}]^T (\underline{\underline{D}}^T \underline{\underline{P}} \; \underline{\underline{D}}) \; (\underline{\underline{D}}^T \underline{\underline{P}} \; \underline{\underline{D}})^{-1} \underline{z} \tag{A/14}$$

With the substitution

$$\underline{w} = (\underline{\underline{D}}^T \underline{\underline{P}} \; \underline{\underline{D}})^{-1} \underline{z} \tag{A/15}$$

we obtain

$$\underline{z}^T (\underline{D}^T \underline{P} \, \underline{D})^{-1} \underline{z} = \underline{w}^T \underline{D}^T \underline{P} \, \underline{D} \, \underline{w}. \tag{A/16}$$

The last expression is positive. It vanishes only for $\underline{w} = \underline{0}$ because of the assumed positive definiteness of the matrix $\underline{D}^T \underline{P} \, \underline{D}$. Because of the regular transformation (A/15) the vector $\underline{w}$ vanishes only in the case $\underline{z} = \underline{0}$. But this means that the matrix $(\underline{D}^T \underline{P} \, \underline{D})^{-1}$ is positive definite.

If the matrix $\underline{D}$ has not the maximal rank n but only a certain rank r with

$$r < n, \tag{A/17}$$

then we have [1]

$$\text{rank} \, (\underline{D}^T \underline{P} \, \underline{D}) = r < n, \tag{A/18}$$

i. e. the matrix $\underline{D}^T \underline{P} \, \underline{D}$ is not invertible.

References:
[1] Zurmühl, R.:
    Matrizen
    4.Auflage
    Berlin, Heidelberg, New York: Springer-Verlag (1964).
[2] Faddejew, D.K.; Faddejewa, W.N.:
    Numerische Methoden der linearen Algebra
    4. Auflage
    München und Wien: R. Oldenbourg Verlag (1976).

Some facts from linear algebra

Here some facts from linear algebra are presented which are necessary for the under-standing of the derivation of the new method for the solution of nonlinear least squares problems. These facts may be found for instance in [1, 2, 3, 4].

For the following considerations a finite dimensional linear vector space is taken as a basis. Let the elements of this linear vector space be m-dimensional vectors with real-valued components. For this linear vector space we write $\mathbf{R}^m$.

A subspace $\mathbb{B}$ of a finite dimensional vector space $\mathbf{R}^m$ is a set of vectors of $\mathbf{R}^m$ such that if $\underline{z}_1$ and $\underline{z}_2$ are two elements in $\mathbb{B}$, then for all real scalars $\alpha$ and $\beta$ also the linear combination $\alpha \underline{z}_1 + \beta \underline{z}_2$ is an element in $\mathbb{B}$.

Let $\mathbb{B}_1$ and $\mathbb{B}_2$ be two subspaces of the vector space $\mathbf{R}^m$. The space $\mathrm{I\!R}^m$ is called the direct sum of the two subspaces $\mathbb{B}_1$ and $\mathbb{B}_2$, if any vector $\underline{z} \in \mathrm{I\!R}^m$ may be written in one and only one way as

$$\underline{z} = \underline{z}_1 + \underline{z}_2 \quad \text{with } \underline{z}_1 \in \mathbb{B}_1 \text{ and } \underline{z}_2 \in \mathbb{B}_2 . \tag{B/1}$$

This fact may be symbolized by

$$\mathbf{R}^m = \mathbb{B}_1 \oplus \mathbb{B}_2 . \tag{B/2}$$

We call the set of all vectors which are orthogonal to every vector in $\mathrm{I\!B}$ the ortho-gonal complement of the subspace $\mathbb{B}$ in the vector space $\mathbf{R}^m$. We write $\mathbb{B}^{\perp}$ for the or-thogonal complement. It is possible to decompose the space $\mathrm{I\!R}^m$ with the help of the subspace $\mathbb{B}$ and its orthogonal complement $\mathbb{B}^{\perp}$ according to

$$\mathbf{R}^m = \mathbb{B} \oplus \mathbb{B}^{\perp} . \tag{B/3}$$

From the decomposition (B/3) we see that any vector $\underline{z}$ can be represented in one and only one way by

$$\underline{z} = \underline{z}" + \underline{z}^{\perp} \tag{B/4}$$

whereby $\underline{z}" \in \mathbb{B}$ and $\underline{z}^{\perp} \in \mathbb{B}^{\perp}$ is valid. The vector $\underline{z}"$ is called the orthogonal project-ion of the vector $\underline{z}$ on the subspace $\mathbb{B}$.

In particular we consider the subspace $\mathbb{B}$ of the vector space $\mathbf{R}^m$ which is spanned by the column vectors of a mxn-matrix $\underline{D}$ with $m \geq n$. This space is called the column space of $\underline{D}$ and is denotated by $R(\underline{D})$. Let $R(\underline{D})^{\perp}$ be the orthogonal complement of $R(\underline{D})$ in the space $\mathbf{R}^m$. Then we find a decomposition on the analogy of eq. (B/3)

$$\mathbf{R}^m = R(\underline{D}) \oplus R(\underline{D})^{\perp} . \tag{B/5}$$

Eq. (B/5) means that any vector $\underline{z} \in \mathbb{R}^m$ can uniquely be represented as the linear combination of the column vectors of the matrix $\underline{D}$ and of vectors which are orthogonal to them. Therefore we can decompose the vector $\underline{z}$ according to

$$\underline{z} = \underline{z}'' + \underline{z}^{\perp} \text{ with } \underline{z}'' \in R(\underline{D}) \text{ and } \underline{z}^{\perp} \in R(\underline{D})^{\perp} . \tag{B/6}$$

$\underline{z}^{\perp}$ is that component of the vector $\underline{z}$ which is orthogonal to all column vectors of the matrix $\underline{D}$.

Sometimes it is interesting to know the conditions for a linear transformation $\underline{T}$ to carry an arbitrary vector $\underline{z} \in \mathbb{R}^m$ into $\underline{z}'' \in R(\underline{D})$ for a given mxn-matrix $\underline{D}$.

The generalized formulation of the question refers to the decomposition (B/1) for an arbitrary vector $\underline{z} \in \mathbb{R}^m$. We will examine the conditions for a linear transformation $\underline{T}$, which carries any vector $\underline{z} \in \mathbb{R}^m$ into $\underline{z}_1 \in \mathbb{B}_1$.

The linear transformation $\underline{T}$ for which

$$\underline{T}\, \underline{z} = \underline{z}_1 \tag{B/7}$$

is valid, is called the projector on $\mathbb{B}_1$ along $\mathbb{B}_2$.

Now the following theorem is valid [3, 4]:

The linear transformation $\underline{T}$ is the projector on $\mathbb{B}_1$ along $\mathbb{B}_2$ if and only if

$$\underline{T}^2 = \underline{T} , \tag{B/8}$$

whereby $\mathbb{B}_1 = \{\underline{z} \mid \underline{T}\, \underline{z} = \underline{z} \}$ and $\mathbb{B}_2 = \{\underline{z} \mid \underline{T}\, \underline{z} = \underline{0} \}$. A linear transformation which satisfies the condition (B/8) is called idempotent.

A projector $\underline{T}$ is called an orthogonal projector if it is a projector with the supplementary property that the vector $\underline{z}_1$ is orthogonal to the vector $\underline{z} - \underline{z}_1$ for any vector $\underline{z}$.

It can be proved that $\underline{T}$ is an orthogonal projector if and only if

$$\underline{T}^2 = \underline{T} \text{ and } \underline{T}^T = \underline{T} \tag{B/9}$$

is valid. This means that $\underline{T}$ has to be idempotent and symmetric [4].

Let the column space $R(\underline{D})$ of a mxn-matrix $\underline{D}$ be the considered subspace of the vector space $\mathbb{R}^m$. If it is demanded that an arbitrary vector $\underline{z} \in \mathbb{R}^m$ is decomposed according to eq. (B/6), so it is necessary that the linear transformation $\underline{T}$ satisfies the conditions (B/9). So it is guaranteed that the vector $\underline{z}$ will be decomposed into two vectors which are orthogonal to each other. Additionally the relation

$$\underline{T}\, \underline{D} = \underline{D} \tag{B/10}$$

must be valid so that the column space of the matrix $\underline{D}$ is transformed into itself.

References:

[1] Boullion, T.L.; Odell, P.L.:
    Generalized Inverse Matrices
    New York: Wiley-Interscience (1971).
[2] v. Mangoldt, H.; Knopp, K.:
    Einführung in die Höhere Mathematik
    Vierter Band, 2. durchgesehene Auflage
    Stuttgart: S. Hirzel Verlag (1975).
[3] Pease, M.C.:
    Methods of Matrix Algebra
    New York and London: Academic Press (1965).
[4] Zadeh, L.A.; Desoer, Ch.A.:
    Linear System Theory
    New York: McGraw-Hill, Inc. (1963).

## Appendix C

The solution of the equation $\underline{D}_{n-1}^k \underline{c}^k = \underline{0}$

The matrix $\underline{D}_{n-1}^k$ is a $m \times (n-1)$-matrix, i.e. it has $n-1$ column vectors. These $n-1$ column vectors can be linearly dependent or linearly independent. It is not known which case is under consideration actually.

Now the following theorem holds [1]:
Given a $m \times (n-1)$-matrix $\underline{D}_{n-1}^k \neq \underline{0}$ and a $(n-1) \times m$-matrix $\underline{c}^k$. The solution of the equation

$$\underline{D}_{n-1}^k \underline{c}^k = \underline{0} \tag{C/1}$$

is

$$\underline{c}^k = \underline{0} \tag{C/2}$$

if and only if the column vectors of the matrix $\underline{D}_{n-1}^k$ are linearly independent.

This means that the choice (C/2) is mandatory if the column vectors of the matrix $\underline{D}_{n-1}^k$ are linearly independent. As the examination of the linear independence can be a numerically difficult problem, it makes sense to choose $\underline{c}^k = \underline{0}$ in the case of the linear independence as well as in the case of the linear dependence of the column vectors. This choice also saves computation time, because it is not necessary to determine the matrix $\underline{c}^k$.

If we have

$$\underline{D}_{n-1}^k = \underline{0} \tag{C/3}$$

we can choose the matrix $\underline{c}^k$ arbitrarily. In this case we again have the problem to decide whether the matrix $\underline{D}_{n-1}^k$ is really equal to the null matrix or not. To avoid this decision the choice (C/2) for the matrix $\underline{c}^k$ is advantageous.

Comprehensively we can state that the decision to set the matrix $\underline{c}^k$ equal to the null matrix is convenient and well-founded.

Reference:

[1] Zurmühl, R.:
    Matrizen
    4. Auflage
    Berlin, Heidelberg, New York: Springer-Verlag (1964).

Appendix D

The solution of the system of linear equations $\underline{D}\,\underline{r} = \underline{e}$

We will consider the system of linear equations

$$\underline{D}\,\underline{r} = \underline{e} \; . \tag{D/1}$$

The matrix $\underline{D}$ is a mxn-matrix with $m \geq n$ and $\underline{e}$ is a vector in the space $IR^m$. It is well-known that such a system of linear equations possesses a solution only in special cases. By solution we mean such a vector $\underline{r}$ which satisfies the system of linear equations (D/1) exactly. If such an exact solution does not exist it is possible to look for a "best" approximate solution of the system of linear equations in the sense that it minimizes the error as measured in an appropriate norm. Here we will consider the solution in the sense of

$$\|\underline{D}\,\underline{r} - \underline{e}\| \leq \|\underline{D}\,\underline{z} - \underline{e}\| \text{ for all } \underline{z} \in \mathbf{R}^n, \tag{D/2}$$

where the norm is the Euclidean norm given by $\|\underline{x}\| = \sqrt{\underline{x}^T \underline{x}}$.

If the solution $\underline{r}$ in the sense of the inequality (D/2) is found, the vector $\underline{D}\,\underline{r}$ is an element in the column space of the matrix $\underline{D}$, i.e. $\underline{D}\,\underline{r} \in R(\underline{D})$. Often it happens that we are not interested in the solution of eq. (D/2) for which the vector $\underline{D}\,\underline{r}$ is an element of the space spanned by all linearly independent column vectors of the matrix $\underline{D}$, but in a solution of eq. (D/2) for which the vector $\underline{D}\,\underline{r}$ is an element of the space spanned only by a subset of the linearly independent column vectors of the matrix $\underline{D}$. We choose l linearly independent column vectors $\underline{d}_1, \ldots, \underline{d}_l$ and build up the matrix $\underline{D}_l$. The set of all vectors in the column space of the matrix $\underline{D}_l$ can be represented as $\underline{D}_l \underline{z}$, whereby $\underline{z}$ is an arbitrary vector from the space $IR^l$. Looking for a solution of the system of linear equations (D/1) in the sense of a best approximate solution with the constraint that the vector $\underline{D}\,\underline{r}$ is an element in the column space of the matrix $\underline{D}_l$ means that we must determine a vector $\underline{r}$ for which is valid

$$\|\underline{D}\,\underline{r} - \underline{e}\| \leq \|\underline{D}_l \underline{z} - \underline{e}\| \text{ for all } \underline{z} \in \mathbf{R}^l$$

$$\text{and } \underline{D}\,\underline{r} \in R(\underline{D}_l) \; . \tag{D/3}$$

For the solution of this problem the following theorem is helpful [1]:
Corresponding to any vector $\underline{e} \in \mathbf{R}^m$ there is a unique vector $\underline{D}\,\underline{r} \in R(\underline{D}_l)$ such that

$$\|\underline{D}\,\underline{r} - \underline{e}\| \leq \|\underline{D}_l \underline{z} - \underline{e}\| \text{ for all } \underline{z} \in \mathbf{R}^l. \tag{D/4}$$

A necessary and sufficient condition for the solution $\underline{D}\,\underline{r}$ of the problem (D/4) is

$$\underline{D}_l \underline{z} \perp (\underline{D}\,\underline{r} - \underline{e}) \text{ for all } \underline{z} \in \mathbf{R}^l, \tag{D/5}$$

this means

$$(\underline{D}\ \underline{r} - \underline{e})^T \underline{D}_1 \underline{z} = 0\ . \tag{D/6}$$

Let the vector $\underline{r}$ be given by

$$\underline{r} = \underline{\bar{R}}\ \underline{e}\ , \tag{D/7}$$

whereby the matrix $\underline{\bar{R}}$ satisfies the following three equations

$$(\underline{D}\ \underline{\bar{R}})^2 = \underline{D}\ \underline{\bar{R}} \tag{D/8}$$

$$(\underline{D}\ \underline{\bar{R}})^T = \underline{D}\ \underline{\bar{R}} \tag{D/9}$$

$$\underline{D}\ \underline{\bar{R}}\ \underline{D}_1 = \underline{D}_1\ . \tag{D/10}$$

These equations correspond to the eqs. (4.2/6), (4.2/7), and (4.7/14) of the new method for the solution of nonlinear least squares problems.

Now we will prove that the vector $\underline{D}\ \underline{r}$ is a solution of the problem (D/3), if the vector $\underline{r}$ is given by eq. (D/7). First of all it is evident because of the eqs. (D/8), (D/9), and (D/10) that the vector $\underline{D}\ \underline{r}$ is an element in the column space of the matrix $\underline{D}_1$. It is sufficient to examine the validity of the relation (D/6)

$$\begin{aligned}
(\underline{D}\ \underline{\bar{R}}\ \underline{e} - \underline{e})^T \underline{D}_1 \underline{z} &= \underline{e}^T(\underline{D}\ \underline{\bar{R}} - \underline{I})\underline{D}_1 \underline{z} \\
&= \underline{e}^T(\underline{D}\ \underline{\bar{R}}\ \underline{D}_1 - \underline{D}_1)\underline{z} \\
&= \underline{e}^T \underline{0}\ \underline{z} = 0\ .
\end{aligned} \tag{D/11}$$

So it has been proved that the solution of the problem (D/3) is given by eq. (D/7) in connexion with the eqs. (D/8), (D/9), and (D/10).

We are to remark that there are several vectors $\underline{r}$ in general which are solution of the problem (D/3), because there are several matrices $\underline{\bar{R}}$ in general which satisfy the eqs. (D/8), (D/9), and (D/10). Because of the above cited theorem it is not the vector $\underline{r}$ which is given uniquely but only the vector $\underline{D}\ \underline{r}$.

Given a vector $\underline{r}$ which is solution of the problem (D/3) it is possible to define an error Q' concerning the solution of the system of the linear equations (D/1). This error is given by

$$\begin{aligned}
Q' &= \|\underline{D}\ \underline{r} - \underline{e}\|^2 = \|\underline{D}\ \underline{\bar{R}}\ \underline{e} - \underline{e}\|^2 \\
&= \|\underline{e}^{\perp}\|^2 \text{ with } \underline{e}^{\perp} \in R(\underline{D}_1)^{\perp}\ .
\end{aligned} \tag{D/12}$$

Because of the above-mentioned remark there are several vectors $\underline{r}$ in general which cause the same error Q'.

One possible solution $\underline{r}$ is given by

$$\underline{r} = \begin{pmatrix} \underline{D}_1^+ \underline{e} \\ \\ \underline{0} \end{pmatrix} = \overline{\overline{R}} \, \underline{e} \; , \tag{D/13}$$

whereby the matrix $\underline{D}_1^+$ is defined by eq. (4.2/21). Namely it has been proved that the matrix $\overline{\overline{R}}$ - defined by eq. (D/13) - satisfies the eqs. (D/8), (D/9), and (D/10) (comp. the eqs. (4.2/28), (4.2/29), and (4.2/30)).

Now we assume that it is possible to determine the rank r of the matrix $\underline{D}$ exactly. Then there are r linearly independent vectors among the column vectors of the matrix $\underline{D}$. The selection of these r linearly independent vectors is not uniquely given in general, if r < n is valid. If we are looking for the solution of the problem (D/2), the vector $\underline{r}$ in eq. (D/13) (setting 1 = r) is not uniquely determined because of the different possibilities in the selection of the column vectors. Therefore we have several solutions $\underline{r}$ in general. All these solutions $\underline{r}$ of the problem (D/2) have in common that at least n-r components of the solution vector $\underline{r}$ vanish. Such a solution $\underline{r}$ is called a basic approximate solution [2]. Besides these solutions with at least n-r vanishing components there is still another or more. We obtain one further solution if we demand that the matrix $\overline{\overline{R}}$ does not only satisfy the eqs. (D/9) and (D/10) but also the equations

$$\overline{\overline{R}} \, \underline{D} \, \overline{\overline{R}} = \overline{\overline{R}} \tag{D/14}$$

and

$$(\overline{\overline{R}} \, \underline{D})^T = \overline{\overline{R}} \, \underline{D} \; . \tag{D/15}$$

(Here we only consider the problem (D/2). In section 4.2 we have proved that we obtain eq. (D/8) from eq. (D/10) if we are able to determine the rank of the matrix $\underline{D}$ exactly. In this case eq. (D/8) is unnecessary.)

At first we must know if such a matrix $\overline{\overline{R}}$ - satisfying eqs. (D/9), (D/10), (D/14), and (D/15) - exists at all. Penrose has proved the existence [3]. Besides the existence he has also proved the uniqueness of this matrix $\overline{\overline{R}}$. This means that we only have one solution of the problem (D/2) when utilizing this matrix $\overline{\overline{R}}$. The matrix $\overline{\overline{R}}$ satisfying the eqs. (D/9), (D/10), (D/14), and (D/15) is called the Moore-Penrose pseudoinverse and is denotated by $\underline{D}^+$. If the column vectors of the matrix $\underline{D}$ are linearly independent, we have the following relation for the matrix $\underline{D}^+$

$$\underline{D}^+ = (\underline{D}^T \underline{D})^{-1} \underline{D}^T \; . \tag{D/16}$$

Utilizing the relation (D/16) for the matrix $\overline{\overline{R}}$ in the eqs. (D/9), (D/10), (D/14), and (D/15) the correctness of the given relation can easily be shown.

Considering eq. (D/16) we see that the notation $\underline{D}_1^+$ for the matrix $(\underline{D}_1^T \underline{D}_1)^{-1} \underline{D}_1^T$ in eq. (4.2/21) is meaningful because we assume that the column vectors of the matrix $\underline{D}_1$ are linearly independent. Therefore the matrix $\underline{D}_1^+$ is the Moore-Penrose pseudoinverse of the matrix $\underline{D}_1$.

Besides the solutions $\underline{r}$ from eq. (D/13) for the problem (D/2) we have found a further solution and that is

$$\underline{r} = \underline{D}^+\underline{e} \ . \tag{D/17}$$

This solution has the property that it is that solution among all possible solutions for which $\|\underline{r}\|$ is minimal [4, 5]. This solution is often utilized in statistical applications [6, 7]. But there are some problems for which the solution with the help of the Moore-Penrose pseudoinverse does not yield the desired result. To see this we consider the example given in [8].

Given the linear mathematical model

$$f = x_1 + x_2 t \tag{D/18}$$

and the two measured values $y_1 = y(t_1)$ and $y_2 = y(t_2)$. It is the problem to determine the two parameters $x_1$ and $x_2$. As long as $t_1 \neq t_2$ is valid, we find

$$\begin{pmatrix} x_1 \\ x_2 \end{pmatrix} = \begin{pmatrix} \dfrac{t_2 y_1 - t_1 y_2}{t_2 - t_1} \\ \dfrac{y_2 - y_1}{t_2 - t_1} \end{pmatrix} \ . \tag{D/19}$$

But now let $t_1 = t_2 = \bar{t}$. By evaluation of eq. (D/17) (methods for the calculation of $\underline{D}^+$ can be found for instance in [4, 6, 9, 10]) we find as solution for the parameters

$$\begin{pmatrix} x_1 \\ x_2 \end{pmatrix} = \begin{pmatrix} \dfrac{y_1 + y_2}{2 \, (1+\bar{t}^2)} \\ \dfrac{\bar{t} \, y_1 + \bar{t} \, y_2}{2 \, (1+\bar{t}^2)} \end{pmatrix} \ . \tag{D/20}$$

In the case $t_1 = t_2 = \bar{t}$ the mathematical model (D/18) is not correct because we cannot distinguish a dependence of the mathematical model on t if we only have the two values $y_1(\bar{t})$ and $y_2(\bar{t})$. In this case we would regard the mathematical model

$$f = x_1 \tag{D/21}$$

as appropriate. The parameter $x_1$ is given by

$$x_1 = \frac{y_1 + y_2}{2} \ . \tag{D/22}$$

The appropriate mathematical model (D/21) with the parameter $x_1$ from eq. (D/22) can be derived in this case from the unsuitable mathematical model (D/18) with the help of eq. (D/13). As we recognize by eq. (D/20) this is not possible if we utilize the Moore-Penrose pseudoinverse for the solution.

This example demonstrates that there are problems for which the solution with the help of the Moore-Penrose pseudoinverse does not absolutely yield the actually desired result.

Now we show that the matrix $\underline{\underline{R}}$ from eq. (D/13) also satisfies eq. (D/14) besides the eqs. (D/8), (D/9), and (D/10). This can be seen by evaluating

$$\underline{\underline{R}}\, \underline{\underline{D}}\, \underline{\underline{R}} = \begin{pmatrix} \underline{D}_1^+ \\ \underline{0} \end{pmatrix} (\underline{D}_1, \underline{D}_{n-1}) \begin{pmatrix} \underline{D}_1^+ \\ \underline{0} \end{pmatrix}$$

$$= \begin{pmatrix} \underline{D}_1^+ \\ \underline{0} \end{pmatrix} (\underline{D}_1 \underline{D}_1^+) = \begin{pmatrix} \underline{D}_1^+ \underline{D}_1 \underline{D}_1^+ \\ \underline{0} \end{pmatrix}$$

$$= \begin{pmatrix} \underline{D}_1^+ \\ \underline{0} \end{pmatrix} = \underline{\underline{R}} \quad . \tag{D/23}$$

Now we especially consider the solution of the problem (D/2) and assume that we are able to determine the rank of the matrix $\underline{D}$ exactly. In this case - i.e. $l = r$ - the matrix $\underline{\underline{R}}$ from eq. (D/13) is called a weak pseudoinverse and denotated by $\underline{D}^{\#}$. It is called a weak pseudoinverse because it satisfies three of the four defining equations of the Moore-Penrose pseudoinverse. The eq. (D/15) cannot also be satisfied because the Moore-Penrose pseudoinverse is determined uniquely. But the weak pseudoinverse can be identical with the Moore-Penrose pseudoinverse. It can be shown that the weak and the Moore-Penrose pseudoinverse are identical if and only if the rank of the matrix $\underline{D}$ is equal to n [2].

References:

[1] Luenberger, D.G.:
    Optimization by Vector Space Methods
    New York: J. Wiley & Sons, Inc. (1969).
[2] Rosen, J.B.:
    Minimum and Basic Solutions to Singular Linear Systems
    SIAM J. Appl. Math. 1, 156-162 (1964).
[3] Greville, T.N.E.:
    The Pseudoinverse of a Rectangular or Singular Matrix and its Application to the
    Solution of Systems of Linear Equations
    SIAM Review 1, 38-43 (1959).
[4] Rao, C.R.; Mitra, S.K.:
    Generalized Inverse of Matrices and its Applications
    New York: J. Wiley & Sons, Inc. (1971).

[5] Zadeh,L.A.; Desoer, Ch.A.:
    Linear System Theory
    New York: McGraw-Hill, Inc. (1963).
[6] Albert, A.:
    Regression and the Moore-Penrose-Pseudoinverse
    New York and London: Academic Press (1972).
[7] Boullion, T.L.; Odell, P.L.:
    Generalized Inverse Matrices
    New York: Wiley-Interscience (1971).
[8] Harris, W.A.; Helvig, T.N.:
    Applications of the Pseudoinverse to Modeling
    Technometrics 2, 351-357 (1966).
[9] Bard, Y.:
    Nonlinear Parameter Estimation
    New York and London: Academic Press (1974).
[10] Greville, T.N.E.:
     Some Applications of the Pseudoinverse of a Matrix
     SIAM Review 1, 15-22 (1960).

## Appendix E

A remark concerning the solution of nonlinear least squares problems with the help of the Moore-Penrose pseudoinverse

The direction vector $\underline{r}^k$ of Hartley's method is given by eq. (3.2/6). If we pay attention to the relations (A/2), (A/5), and (4.2/2) it can be written as

$$\underline{r}^k = ((\underline{D}^k)^T \underline{D}^k)^{-1} (\underline{D}^k)^T \underline{e}(\underline{x}^k) . \tag{E/1}$$

In order to apply eq. (E/1) we must assume that the column vectors of the matrix $\underline{D}^k$ are linearly independent. As shown in appendix D, we can interpret the direction vector $\underline{r}^k$ as solution of the system of linear equations

$$\underline{D}^k \underline{r}^k = \underline{e}(\underline{x}^k) . \tag{E/2}$$

Looking at this system of linear equations we recognize immediately how to extend Hartley's method in the case that the column vectors of the matrix $\underline{D}^k$ are linearly dependent [1].

As direction vector $\underline{r}^k$ we choose the solution of the system of linear equations (E/2) in the sense of the relation (D/2). As shown in appendix D, this solution always exists and it is given by

$$\underline{r}^k = \underline{\bar{R}}^k \underline{e}(\underline{x}^k) \tag{E/3}$$

whereby the matrix $\underline{\bar{R}}^k$ satisfies the following four equations

$$\underline{D}^k \underline{\bar{R}}^k \underline{D}^k = \underline{D}^k \tag{E/4}$$

$$(\underline{D}^k \underline{\bar{R}}^k)^T = \underline{D}^k \underline{\bar{R}}^k \tag{E/5}$$

$$\underline{\bar{R}}^k \underline{D}^k \underline{\bar{R}}^k = \underline{\bar{R}}^k \tag{E/6}$$

$$(\underline{\bar{R}}^k \underline{D}^k)^T = \underline{\bar{R}}^k \underline{D}^k . \tag{E/7}$$

The matrix $\underline{\bar{R}}^k$ defined by these four equations is the Moore-Penrose pseudoinverse. If we determine the direction vector $\underline{r}^k$ according to eq. (E/3), the so modified Hartley's method is always applicable. For its application it is necessary to determine the matrix $\underline{\bar{R}}^k$ by any means. That means for the processing of the method on a digital computer that we must determine the matrix $\underline{\bar{R}}^k$ with the help of a numerical method. As mentioned in connexion with the derivation of the new method for the solution of nonlinear least squares problems recursive methods are well suitable for this problem. For many of these methods we have to decide in their course if a vector $\underline{d}^k_{i+1}$ to be selected now is linearly dependent on the already selected vectors $\underline{d}^k_1, \ldots, \underline{d}^k_i$. Concerning the new method for the solution of nonlinear least squares problems this question is answered with the help of the scalar $a^k_i$ (see eq. (4.3/13)).

Greville presents a recursive method for the determination of the Moore-Penrose pseudoinverse [2]. He also utilizes the scalar $a_1^k$ for the necessary decision. Applying Greville's method on a digital computer, the problem occurs when to consider the scalar $a_1^k$ as zero.

For the illustration of the possible difficulties in connexion with the method  given by Greville we consider the following problem. Assume that n-1 vectors are found to be linearly independent and that the corresponding Moore-Penrose pseudoinverse is determined. Now we add the last vector $\underline{d}_n^k$. This vector is linearly independent on the other vectors, if we perform the examination analytically. If we perform this examination on a digital computer with the help of the scalar $a_1^k$ we find that the vector $\underline{d}_n^k$ is linearly dependent on the other vectors, i. e. the scalar $a_1^k$ is less than a given upper bound $\varepsilon > 0$. Therefore the Moore-Penrose pseudoinverse - calculated on a digital computer - will be different from the Moore-Penrose pseudoinverse $\underline{R}^k$ determined analytically. This difference would be tolerable if the stability of  the resulting method is guaranteed. But it turns out that in general the matrix $\underline{D}^k\underline{R}^k$ is no longer an orthogonal projector but only a projector. This is evident if for the above described case we determine the Moore-Penrose pseudoinverse with the help of the algorithm given by Greville and then examine the conditions (B/9) for an orthogonal projector.

But for the guarantee of the stability of the described method the fact is important that the matrix $\underline{D}^k\underline{R}^k$ is an orthogonal projector (comp. eq. (4.2/11)). Naturally this does not mean that the method for the solution of nonlinear least squares problems with the help of the Moore-Penrose pseudoinverse - calculated by Greville's method - will automatically be unstable if a linearly independent vector - analytically examined - is considered as linearly dependent for the further processing on a digital computer. But nevertheless, there is the possibility that the sufficient condition for the stability of the method is not satisfied. Such an uncertainty should be avoided.

Similar investigations can be made if not only one vector is considered as linearly dependent but several. It is always found that the sufficient condition for the stability of the method is not satisfied in general.

For this reason the direction vector $\underline{r}^k$ given by eq. (E/3) with the help of the Moore-Penrose pseudoinverse is in general unsuitable for the solution of nonlinear least squares problems, if we calculate $\underline{r}^k$ with the help of Greville's method.

Another possibility for the determination of the direction vector $\underline{r}^k$ given by eq. (E/3) stems from the use of the singular value decomposition of the matrix $\underline{D}^k$ (compare appendix F). With its help it is easy to calculate the Moore-Penrose pseudoinverse of $\underline{D}^k$, that means the matrix $\underline{R}^k$ satisfying the eqs. (E/4 - 7). Applying the singular value decomposition it can happen that we have to consider one or more singular values $\sigma_j$ as zero if we perform their calculation on a digital computer although they are different from zero when performing their calculation analytically. In this case we obtain a matrix $\bar{R}^k$ which is different from the corresponding analytically determined matrix. But we can show that this matrix $\bar{R}^k$ has the property that the matrix $\underline{D}^k\bar{R}^k$ is an orthogonal projector in the space spanned by the vectors $\underline{d}_j^k$ associated with those singular values which are different from zero respectively which are not considered as zero. So the stability of the resulting method is guaranteed. But in this case it can happen that the scalar product $(\underline{\nabla} Q(\underline{x}^k))^T \underline{r}^k$ vanishes although the gradient of the sum of squares differs from the null vector. This means that we have no decrease of the sum of squares in the direction of the so determined vector $\underline{r}^k$ at the point $\underline{x}^k$ although the necessary condition for an extremum is not satisfied. This is an undesired situation. Because of this drawback we can state that the determination of the direction vector $\underline{r}^k$ with the help of the Moore-Penrose pseudoinverse by utilizing the singular value decomposition is not well appropriate to the solution of nonlinear least squares problems. But if we want to bear this disadvantage the singular value decomposition is one possibility for the determination of the direction vector $\underline{r}^k$ which also yields the chance to influence the condition of the system of linear equations to be solved. Inspecting eq. (E/2) we see that $\underline{r}^k$ is the solution of a system of linear equations. In this connexion the condition of the matrix $\underline{D}^k$ is of interest. A measure for the condition is the condition number $\varkappa$ which is given by the ratio of the greatest to the smallest singular value $\sigma_j$ of the matrix $\underline{D}^k$ [3]. As we can consider one or more singular values as zero we can achieve that the condition number $\varkappa$ does not exceed a given upper bound. But we must emphasize once more that this processing may yield a direction vector $\underline{r}^k$ which does not guarantee a decrease of the sum of squares although its gradient is not equal to the null vector.

References:

[1] Fletcher, R.:
    Generalized Inverse Method for the Best Least Squares Solutions of Systems of
    Nonlinear Equations
    The Computer Journal 10, 392-399 (1967).
[2] Greville, T.N.E.:
    Some Applications of the Pseudoinverse of a Matrix
    SIAM Review 1, 15-22 (1960).
[3] Klema, V.C.; Laub,A.J.:
    The Singular Value Decomposition: Its Computation and Some Applications
    IEEE Trans. Autom. Control AC-25, 164-176 (1980).

## Appendix F

The singular value decomposition

Given a real mxn-matrix $\underline{A}$ (m≥n) with rank $\underline{A}$ = r. Then there exist a real orthogonal mxn-matrix $\underline{U}$, a real orthogonal nxn-matrix $\underline{V}$, and a real diagonal matrix $\underline{\Sigma}$ such that

$$\underline{A} = \underline{U} \, \underline{\Sigma} \, \underline{V}^T \tag{F/1}$$

where

$$\underline{\Sigma} = \text{diag}(\sigma_1, \ldots, \sigma_n) . \tag{F/2}$$

(For a proof see for instance [1].)

The scalars $\sigma_j$ are non-negative. We assume that they are numbered in such a way that

$$\sigma_1 \geq \sigma_2 \geq \ldots \sigma_n \geq 0 \tag{F/3}$$

holds. Because of the assumption that the matrix $\underline{A}$ is of rank r, the scalars $\sigma_1$, ..., $\sigma_r$ are positive while the other scalars $\sigma_{r+1}$, ..., $\sigma_n$ vanish that means

$$\sigma_{r+1} = \ldots = \sigma_n = 0 \tag{F/4}$$

holds. The scalars $\sigma_1^2$, ..., $\sigma_n^2$ are the eigenvalues of the matrix $\underline{A}^T\underline{A}$. They are real and non-negative because the matrix $\underline{A}^T\underline{A}$ is symmetric and at least positive semide-finite. The non-negative square roots of these eigenvalues are called the singular values $\sigma_j$ of the matrix $\underline{A}$. The column vectors of the matrix $\underline{U}$ are called the left singular vectors of the matrix $\underline{A}$. They are the n orthonormal eigenvectors of the matrix $\underline{A} \, \underline{A}^T$ associated with the n largest eigenvalues of this matrix, that means

$$\underline{A} \, \underline{A}^T\underline{U} = \underline{U} \, \underline{\Sigma}^2 \tag{F/5}$$

with

$$\underline{U}^T\underline{U} = \underline{I}_n \tag{F/6}$$

holds. ($\underline{I}_n$ is the nxn identity matrix.)

The column vectors of the matrix $\underline{V}$ are called the right singular vectors of the matrix $\underline{A}$. They are the orthonormal eigenvectors of the matrix $\underline{A}^T\underline{A}$, that means

$$\underline{A}^T\underline{A} \, \underline{V} = \underline{V} \, \underline{\Sigma}^2 \tag{F/7}$$

with

$$\underline{V}^T\underline{V} = \underline{V} \, \underline{V}^T = \underline{I}_n \tag{F/8}$$

holds. The decomposition of the matrix $\underline{A}$ according to (F/1) is called the singular value decomposition (SVD). By the definition of the singular values we see that the number of positive singular values determines the rank of the matrix $\underline{A}$. From a numerical point of view we have the problem to decide whether a singular value is considered "zero" or not. Therefore we have to introduce a positive constant as zero threshold. If a singular value is less than this constant it is considered zero. Now, it is possible to derive a relation for this constant which depends in the main on the precision of the utilized digital computer [2]. Therefore we can state that the singular value decomposition is the only generally reliable method for the determination of the rank of a given matrix $\underline{A}$ which is known up to date. Clearly the numerical rank of a matrix $\underline{A}$ depends on the chosen constant.

With the help of the singular value decomposition it is possible to derive a relation for the Moore-Penrose pseudoinverse $\underline{A}^+$ of the matrix $\underline{A}$. It is given by

$$\underline{A}^+ = \underline{V}\ \underline{\Sigma}^+\underline{U}^T \tag{F/9}$$

where

$$\underline{\Sigma}^+ = \mathrm{diag}(\sigma_j^+) \tag{F/10}$$

with

$$\sigma_j^+ = \begin{cases} 1/\sigma_j & \text{for } \sigma_j > 0 \\ 0 & \text{for } \sigma_j = 0 \end{cases} \tag{F/11}$$

The correctness of the relation (F/9) is easily established by verifying that the matrix $\underline{A}^+$ from eq. (F/9) satisfies the four equations (E/4-7) which define the Moore-Penrose pseudoinverse.

In order to obtain the singular value decomposition we have to determine the matrices $\underline{U}$ and $\underline{V}$ and the singular values $\sigma_j$. In order to compute the singular values it is not a good idea to find the eigenvalues of the matrix $\underline{A}^T\underline{A}$ because the computation of this matrix introduces errors because of the finite precision of the digital computer. (For an example see for instance [2].) Moreover it is not intelligent to determine the matrix $\underline{V}$ by calculating the eigenvectors of the matrix $\underline{A}^T\underline{A}$ directly because this matrix is usually ill-conditioned. Fortunately, there exists an efficient and stable algorithm which does not show these disadvantages. Such an appropriate numerical method is described in [3]. This method consists of two phases. In phase I the given matrix $\underline{A}$ is reduced to bidiagonal form by the aid of Householder transformations. In phase II the singular values of the bidiagonal matrix produced in phase I are computed with the help of the QR algorithm. (The principles of the Householder transformations and the QR algorithm are described for instance in [1].)

References:

[1] Stewart, G.W.:
    Introduction to Matrix Computations
    New York and London: Academic Press (1973).

[2] Klema, V.C.; Laub, A.J.:
The Singular Value Decomposition: Its Computation and Some Applications
IEEE Trans. Autom. Control AC-25, 164-176 (1980).
[3] Golub, G.H.; Reinsch, C.:
Singular Value Decomposition and Least Squares Solutions
Numer. Math. 14, 403-420 (1970).

# Lecture Notes in Control and Information Sciences

Edited by A. V. Balakrishnan and M. Thoma